Packaging Machinery: Sustainability and Competitiveness

by Padraic J. Sweeney

INTERNATIONAL
TRADE
ADMINISTRATION

Published September 2010.

Prepared under the auspices of the Office of Transportation and
Machinery and the Office of Trade Policy Analysis in the
International Trade Administration's Manufacturing and Services Division.
www.trade.gov

Contents

Preface

About This Series of Reports

The Department of Commerce's "Sustainable Manufacturing Initiative (SMI) Sector Focus Study Series" aims to inform public and private sector stakeholders about the specific sustainability-related challenges, present-day best practices, and unrealized opportunities that exist in specific U.S. manufacturing sectors. By shedding light on the market drivers for an industry sector's natural resource efficiency, the department aims to provide clarity on (a) the specific hurdles U.S. firms are facing in their efforts to become more resource efficient and thus more competitive, (b) what firms are doing to overcome these hurdles, (c) the potential cost-saving and value-adding opportunities associated with the sustainable production practices specific to a selected sector, (d) U.S. government programs and resources designed to help firms in a selected sector meet their sustainability-related goals, and (e) unexplored areas of public-private collaboration that could help enhance the sustainability and competitiveness of U.S. firms in a selected sector.

This paper cites several Web sites of public sector programs and resources designed to support U.S. firms in their sustainable business efforts. For comprehensive access to federal government programs and resources pertaining to sustainability-related issues highlighted in this study, we recommend that readers refer to the Department of Commerce's Sustainable Business Clearinghouse on the department's Sustainable Manufacturing Initiative home page at *www.manufacturing.gov/sustainability*.

The Department of Commerce also welcomes public comments and feedback on this study. Please direct any comments to Padraic Sweeney in the Office of Transportation and Machinery at *padraic.sweeney@trade.gov*, or by phone at (202) 482-5024.

Definitions

The terms "sustainability" and "sustainable manufacturing" are used numerous times throughout this paper. Though a variety of definitions for these terms exist today, for the purposes of this paper, both these terms will refer to manufacturing processes that minimize negative environmental impacts; conserve energy and natural resources; are safe for communities, workers, and consumers; and are economically sound.

"Competitiveness" may be defined as a company's ability to provide goods and services at least as effectively and efficiently, if not more so, than the relevant competitors. Measures of competitiveness include profitability, the extent to which a firm exports, and market share in domestic and international markets.[1]

Sustainability is also referred to frequently in terms of the "triple bottom line" of economic, environmental, and social performance U.S. manufacturers, including packaging machinery OEMs, meet very high workplace safety and other social criteria when compared with many of their overseas competitors. However, this study focuses primarily on the relationship between economic and environmental sustainability.

Abbreviations

ACL	applied ceramic labeling
CE	Conformité Européene (European Conformity)
CPG	consumer packaged goods
DfE	Design for Environment
EPR	extended producer responsibility
EVA	ethylene vinyl acetate
GHG	greenhouse gases
GRI	Global Reporting Initiative
ISO	International Organization for Standards
LCA	lifecycle assessment
NAICS	North American Industry Classification System
OEE	overall equipment effectiveness
OEM	original equipment manufacturer
PMMI	Packaging Machinery Manufacturers Institute
RoHS	Restriction of Hazardous Substances (EU directive)
SMI	Sustainable Manufacturing Initiative
SPC	Sustainable Packaging Coalition
TCO	total cost of ownership
WEEE	Waste Electrical and Electronic Equipment (EU directive)

Executive Summary

U.S. manufacturers of packaging machinery can compete successfully in both domestic and international markets by pursuing business strategies based on sustainability. Many innovative U.S. original equipment manufacturers (OEMs) of packaging machinery are already doing this. The sustainability strategies identified in this report enable U.S. packaging machinery OEMs to target the largest cost per value component of the global packaging market: packaging materials, which are worth an estimated $475 billion annually.[2]

The principal findings of this study include the following:

- Packaging machinery OEMs operate in a global packaging supply chain that faces increasing demands for sustainability.

- Retailers, in particular, play a key role in driving demand for more sustainable packaging throughout the supply chain, even though they generally are not end users of packaging machinery.

- Reducing customers' consumption of packaging materials and ancillary products is the common objective of packaging machinery OEMs that have incorporated sustainability into their core business strategy.

- Reducing customers' packaging-related consumption of energy and water and emissions of greenhouse gases (GHGs) are also key components of successful sustainability strategies.

- Opportunity and innovation drive a successful business strategy based on sustainability for packaging machinery OEMs.

- OEMs with sustainability strategies frequently identify and pursue opportunities for innovation as a result of their ongoing roles as technology suppliers to their customers.

- There is no appreciable demand at present for packaging machinery with sustainable characteristics, as such; end users' procurement practices for packaging machinery do not yet reflect senior management's emphasis on sustainability.

- OEMs are likely to begin encountering demand for packaging machinery with sustainable characteristics in the near future, as their customers aggressively seek to reduce energy and water use, GHG emissions, and waste throughout their manufacturing operations.

- The lack of definitions, certifications, or standards for sustainability in packaging machinery appears to contribute to the lack of demand.

- European laws, regulations, and standards concerning packaging and machinery are shaping the world market.

- Each OEM identified in this study has its own distinctive approach to sustainability, but all of them focus their efforts on technologies and services to reduce customers' consumption of the following:

  - Packaging materials

  - Ancillary products, especially inks and adhesives

  - Energy and water in selected applications

- The cost savings that a focus on materials offers manufacturers of consumer packaged goods (CPG) are what make these OEMs and their products highly competitive.

- Sustainability strategies in the packaging machinery industry typically are oriented around one or more of the following:

  - Automation and integration services and technologies, including remote monitoring

  - Reduction of energy consumption connected with ancillary products

  - Development of innovative ancillary products

  - Development of new packaging systems

- OEMs in this study use one of several recognized methodologies to measure the benefits conferred by their sustainability strategies. These include Life Cycle Assessment (LCA), Total Cost of Ownership (TCO), or Overall Equipment Effectiveness (OEE).

- OEMs in this study have frequently formed strategic relationships with converters or other suppliers of packaging materials or ancillary products.

OEMs of all sizes, involving a variety of business models, are enjoying competitive success with business strategies based on sustainability. In doing so, they are aligning themselves with many others in the packaging supply chain that have also embraced sustainability, including many of their customers. They are also preparing for the day when end users begin demanding more sustainable packaging machinery.

Introduction

*"Waste is something I purchased
but didn't use."*

—Attributed to Henry Ford

The U.S. Department of Commerce has undertaken this study, "Packaging Machinery: Sustainability and Competitiveness," to determine whether U.S. packaging machinery OEMs can implement sustainable business practices and still remain or become more competitive. Although it stands to reason that sustainability contributes to competitiveness—by reducing costs associated with environmental waste—this study attempts to more thoroughly answer that question in a more rigorous manner. As the question was pursued, an important corollary emerged: In practice, what does it mean for packaging machinery OEMs to be sustainable and competitive?

Sustainability can be good for business, even in difficult economic times. A 2009 study by A.T. Kearney found that companies committed to pursuing sustainability achieved above-average financial performance during the recession. Between May and November 2008, providers of industrial goods and services listed on either the Dow Jones Sustainability Index or the Goldman Sachs SUSTAIN focus list outperformed their industry peers by 23 percent. The study linked those sustainability leaders to a variety of sound business practices, including a focus on the long-term well-being of the business, strong corporate governance, sound risk management, and a history of investment in environmental innovation.[3]

In fact, the report "Packaging Machinery: Sustainability and Competitiveness" found that several innovative U.S. packaging machinery OEMs are pursuing business practices based on sustainability and that the practices appear to significantly enhance their competitiveness. This report will identify several of those companies, place them in the context of the packaging machinery industry and the larger packaging supply chain, and describe how their pursuit of sustainability has helped them be more competitive.

Considerable scope exists for making the packaging supply chain at large more sustainable. Globally, large amounts of raw materials are consumed to produce packaging, most of which becomes waste shortly after the goods are purchased. For example, an estimated 30 percent of municipal solid waste in the United States results from discarded packaging of all types.[4] Packaging is also very conspicuous as waste, even though it generally represents only a modest fraction of the overall environmental impact of most packaged consumer products. Finally, sustainability is a relative term with respect to packaging, which involves significant environmental impacts throughout its life cycle. In practice, making packaging more sustainable means mitigating—not eliminating—those impacts.

Strong market and regulatory forces are already at work pushing the global packaging industry toward greater sustainability. Retailers and CPG manufacturers recognize that significant savings can be realized by reducing costs associated with packaging-related wastes. Consumers exert a strong—if not always consistent—influence on retailers and CPG manufacturers as well, through their increasing preference for products that they perceive as environmentally friendly. A growing body of European law, regulation, and standards governing packaging and packaging waste is also shaping the global business environment for packaging goods and services—far beyond the member states of the European Union.

"Packaging Machinery: Sustainability and Competitiveness" was written for two audiences: U.S. packaging machinery manufacturers, their customers, and suppliers and non-packaging specialists with a serious interest in sustainable manufacturing. For U.S. packaging machinery manufacturers, in particular, this study is intended to help them be more competitive and successful in a rapidly changing industry. For those outside the packaging industry, this study is intended to provide some insight into the opportunities and challenges sustainability presents for capital equipment manufacturers in general. As a result, this study attempts to explain a rather specialized topic in language accessible to the specialist and non-specialist alike. Inevitably, some sections of the study will be of greater interest to one audience than to the other.

Because packaging machinery is such a specialized industry, "Packaging Machinery: Sustainability and Competitiveness" begins in section I, "The U.S. Packaging Machinery Industry: Scope and Market Characteristics," with a description of what constitutes packaging machinery, as well as some basic information on packaging materials and the various functions packaging performs. Section II, "The Changing Business Environment for Packaging Machinery," discusses the market and regulatory forces shaping global demand for more sustainable packaging. The core findings relating to packaging machinery OEMs are found in sections III ("Sustainability as Competitive Advantage"); IV ("Manufacturer Case Studies"); and V ("Challenges to Implementing Sustainability").

Research

To produce this report, numerous participants in the packaging machinery industry and the larger packaging supply chain were consulted. Participants included representatives from several packaging machinery OEMs; packaging materials converters; and other market participants in Illinois, Indiana, Minnesota, and Wisconsin. Purdue University Calumet's Department of Mechatronics and the Packaging Machinery Manufacturers Institute

(PMMI) organized a roundtable discussion with several packaging machinery OEMs in Hammond, Indiana. The Sustainable Packaging Coalition (SPC) 2009 spring and fall meetings, the 2009 Sustainable Packaging Forum, and PACK EXPO 2009 all provided valuable opportunities to meet with companies from throughout the packaging supply chain, including numerous CPG manufacturers. Also, an extensive review of available publications on packaging machinery, sustainable packaging, and related topics was conducted.

"Packaging Machinery: Sustainability and Competitiveness" contains several case studies of individual companies' experiences developing and commercializing sustainable products and services. Because a principal objective of this study is to help U.S. packaging machinery OEMs to be more competitive, real-world private-sector examples are given. Accordingly, the mention of any company, product, or service should be viewed as purely illustrative—not as a recommendation or endorsement. Packaging machinery end users looking for specific packaging solutions need to conduct their own thorough due diligence to determine which vendors, products, or services best meet their needs.

Acknowledgments

A number of organizations and individuals were of great assistance in conducting this study. For introductions to packaging machinery OEMs, assistance in organizing roundtable discussions, and general guidance and insight, the author would like to thank Ben Miyares, Matthew Croson, Thomas Egan, and Jorge Izquierdo, vice presidents for industry relations, member services and communications, industry services, and market development of PMMI, respectively (Miyares is now an independent packaging market analyst); Anne Johnson, executive director of SPC; Martha Stephenson, senior project manager at GreenBlue; and Lash Mapa, professor of industrial engineering technology at Purdue University Calumet.

From the companies featured as case studies, the author would like to thank Dale Andersen, president, and Kenneth M. Sullivan, director of marketing, Delkor Systems, Inc.; Scott Smith, director, Global Market Development & Emerging Businesses, Hartness International, an ITW company; Rick Pallante, marketing development manager, Nordson Corporation's packaging adhesives division; Jack

Aguero, vice president for business development and marketing, Pro Mach, Inc., William Chu, general manager, Wexxar Packaging, Inc., a subsidiary of Pro Mach, Inc.; and Paul Appelbaum, president and CEO of Partner Pak, Inc.

The author gratefully acknowledges the assistance and insight provided by Ed Sanders, director of research and development, Food Packaging Americas; Brian K. Muehl, manager of materials technology, Alcan Packaging; Nick Wilson, president, Morrison Container Handling Systems; John Kowal, market development manager, B&R Industrial Automation; Richard Ryan, chief operating officer, Dorner Manufacturing Corp.; Bruce Larson, director of sales, engineered systems, Pearson Packaging Systems; John Naunas, director of sales, Shuttleworth, Inc.; Dana Luthy, director of packaging, Target, Inc.; and Randy L. Spahr, executive vice president, Z Automation Company, Inc.

The author also would like to recognize and express appreciation for the time and guidance so generously provided by employees of the ITW Corporation and several of its subsidiaries, including Kenneth A. Hoffman, group president, ITW Packaging Systems; Robert A. Hank, director of environmental health and safety, and Renita L. Dixon, sustainability and environmental coordinator, Illinois Tool Works, Inc.; Jeff Neitzel, director of marketing, and Al Fernandez, project manager, Hi-Cone; Mark Hughes, manager, applications development and research, and Caitlin A. Rowlands, packaging research engineer, Signode; Kevin D. O'Leary, vice president and group general manager, and Bob Stolmeier, business development manager, Zip-Pak.

For their insights and comments as outside reviewers, the author would like to recognize and thank Todd Bukowski, associate, Packaging & Technology Integrated Solutions, LLC; Tom Egan and Jorge Izquierdo, PMMI; Bruce Larson, Pearson Packaging Systems; and Anne Johnson, SPC.

Finally, the author would be remiss in not recognizing his colleague, William McElnea, international economist in the Commerce Department's Office of Trade Policy Analysis, for his many contributions—conceptual, intellectual, and editorial—to the creation of this document.

I. The U.S. Packaging Machinery Industry: Scope and Market Characteristics

Packaging machinery performs a variety of functions that include canning; container cleaning, filling, and forming; bagging, packing, unpacking, bottling, sealing, and lidding; inspection and check weighing; wrapping, shrink film, and heat sealing; case forming, labeling, and encoding; palletizing and depalletizing; and related applications. Economic data describing the packaging machinery industry is the subject of North American Industry Classification System (NAICS) category 333993, "Packaging Machinery Manufacturing."[5] Sections HS 842220, HS 842230, and HS 842240 of the Harmonized Tariff Schedule of the United States describe U.S. international trade data for packaging machinery.[6] In practice, packaging machinery also includes certain types of materials handling equipment, such as conveyors and accumulators, and specialized printing and graphics machinery.

Packaging machinery manufacturers provide essential technology for a large and increasingly globalized packaging supply chain. Upstream, this supply chain includes producers of basic materials, such as paper, plastic resins, and metals, and packaging materials converters (firms that produce packaging materials from these basic products). Downstream, the supply chain includes CPG manufacturers that package their own products and contract packaging firms that package goods manufactured by other firms.

U.S. packaging machinery manufacturers follow a number of business models. Several larger companies have emerged as providers of complete, integrated turn-key packaging lines. Such companies offer value-added design, engineering, and integration services, along with machinery and traditional after-sales service and support. Others dominate specialized technologies, such as equipment for dispensing adhesives or coding packages. A number of converters also manufacture equipment to process the materials that are their principal business. Many other companies offer specific equipment types, components, and technology services.

The Market for Packaging Machinery

The total U.S. market for packaging machinery in 2008 was worth $6.3 billion, with domestic manufacturers reporting $4.8 billion in sales.[7] The U.S. Census Bureau reports that 551 companies manufactured packaging machinery in the United States in 2007. Most packaging machinery producers are quite small, with nearly 64 percent having fewer than 20 employees.[8]

Manufacturers of processed food and beverages represent approximately 55 percent of the packaging machinery market. Pharmaceutical manufacturers purchase another 10 percent. Other significant packaging machinery end-user segments account for another 20 percent and include household, agricultural, and industrial chemicals; personal care products; hardware; and paper products.[9] Although retailers are not usually end users of packaging machinery, they exert powerful influence over the packaging industry through their purchasing power and increasing focus on more sustainable packaging.

Manufacturers of packaging machinery face a rapidly changing and highly competitive environment. The large CPG manufacturers that purchase most packaging machinery have global supply chains not only for their production inputs, but also for the machinery and materials they use to package their finished goods. Machinery manufacturers face a growing tension between their customers' demand for more flexible, productive equipment and their own need to maintain their profit margins, standards, and reputation for quality.

The leading competitors for U.S. packaging machinery OEMs are, for the most part, European. U.S. industry participants identify European companies' ability to provide turn-key service—design, engineering, and installation of complete processing and packaging lines, rather than simply providing individual machines—as their most important competitive advantage. Leading competitors are from Germany, Italy, and several smaller northern European countries. Japanese manufacturers are also major, well-established competitors. Imports from China have grown strongly in recent years.

The U.S. packaging machinery industry includes many successful exporters that do business with customers around the world. Nevertheless, the industry has lost ground in recent years to foreign competitors. Exports worth $787.4 million represented 14 percent of total shipments in 2007, a slight decrease from 15.1 percent in 2002. Imports worth $2.2 billion accounted for 39 percent of the domestic market the same year, which was up from 26.2 percent in 2002.[10]

Not surprisingly, the recession has affected both U.S. exports and imports of packaging machinery. Exports and imports of packaging machinery both peaked in 2008, at $863.2 million and $2.3 billion, respectively. In 2009, exports fell 16.4 percent to $721.8 million, and imports fell 28.8 percent to $1.6 billion.[11]

Packaging Types and Materials

Packaging can be separated into four basic categories. Packaging machinery is sometimes described according to these categories, as well. Primary packaging directly wraps or contains the product, for example a bottle. Secondary packaging wraps or contains the primary packaging, for example, a plastic wrap containing a small number of bottles. Distribution packaging wraps or contains a product during distribution and provides for efficient handling, for example, a case containing a larger number of bottles. Unit load or transport packaging assembles multiple containers into a single combined bundle suitable for materials handling equipment. For transport, such packaging is frequently stabilized through the use of pallets, strapping, shrink-wrapping, or similar means to form a single unit.[12]

Packaging performs a variety of functions. Packaging protects products during transportation and storage from physical impact, crushing, abrasion, heat, cold, moisture, and other threats that could render the goods unfit for sale. Packaging also protects products from contaminants during transportation and storage, keeping them sanitary and sterile until they are consumed. Packaging contains products so that they can be transported and stored. Packaging provides security from theft and tampering and communicates essential information about products. For CPG manufacturers, packaging also plays a vital role in marketing and establishing brand awareness in an intensely competitive marketplace.[13]

A wide range of materials are used as packaging and processed by packaging machinery. Commonly used materials include paper and paperboard, plastics (rigid, flexible, and films), metals (steel, aluminum, and tin), glass, wood, and textiles. In recent years, paper and paperboard have represented approximately 45 percent of total packaging materials sales, plastics 22 percent, metals nearly 17 percent, and glass and wood slightly more than 4 percent each. Consumer products account for 80 percent of all packaging, including food, beverages, household chemicals, personal care products, and consumer durables such as household appliances, furniture, and computers. Industrial products, such as electrical machinery, medical devices, and other goods account for the balance.[14]

Packaging is a major consumer of materials. For example, approximately 72 percent of converted paperboard, 20 percent of glass, and 18 percent of aluminum are used for packaging. Packaging is a major end use for many ancillary products, including adhesives (44 percent) and ink (32 percent). Packaging is the third-largest market for steel after transportation and construction.

Packaging comes in many forms. Rigid packaging includes containers such as boxes, bottles, drums, cartons, crates, tubs, and pails. Flexible bags, pouches, tubes, wraps, and laminates made of paper, plastic films, and aluminum foil—often in combination (for example, a potato chip bag) are widely used packaging forms. Packaging also includes numerous components and ancillary products, such as closures, tamper-evident materials, cordage, twine, strapping, pallets, skids, and more.[15]

II. The Changing Business Environment for Packaging Machinery

Packaging machinery manufacturers do business in an environment where reducing the overall volume of packaging materials consumed is both a major market demand and, especially in the EU, a legal and regulatory requirement. OEMs must adapt to the fact that sustainability has become a powerful design criteria for new materials and packaging systems that will be run on their machines. Manufacturers must also ensure that their machinery can run more conventional materials whose characteristics are changing because of higher recycled material content.

Three forces are driving the packaging supply chain toward greater sustainability: cost reduction, consumer attitudes, and regulation. Major retailers that purchase most packaged consumer goods increasingly demand that their suppliers reduce the costs associated with packaging and packaging waste—principally by redesigning their packaging to reduce its weight and volume. Eliminating waste at the source rather than after it has been created is commonly referred to as source reduction. The preferences of consumers, a growing share of whom want products they perceive as environmentally friendly, also influence retailer behavior. Many countries, especially in Europe, regulate packaging and packaging waste. It is likely that there will be greater regulation of packaging waste in the United States in the future, as well, especially at the state level.

These forces represent not only necessity, but opportunity, for any company in the packaging supply chain that can capture value by reducing the costs and wastes associated with packaging.

Cost Reduction

CPG manufacturers consume a wide range of packaging materials and generate significant volumes and varieties of waste. Extracting raw materials, converting them into packaging materials, packaging consumer and other products, and transporting both the materials and the packaged goods entail significant costs. Materials wasted during packaging operations, when packages fail before being opened or at other points in a package's life cycle, also represent significant costs. In addition to packaging materials themselves, packaging-related inputs include: hazardous materials, especially petroleum-based resins used in many adhesives and heavy metals contained in many inks; energy, consumed during materials extraction, manufacturing, and conversion, and during packaging and transport operations; and water, as a process input and as a lubricant for bottle, jar, and canning lines. Packaging wastes include discarded packaging materials, greenhouse gases, hazardous wastes, and wastewater.[16]

Packaging inputs that do not result in a saleable product, or are discarded once the good is sold, are waste. Eliminating packaging wastes before they are created—source reduction—can lower manufacturing costs for companies throughout the packaging supply chain. When processed efficiently, recovered packaging waste can also return significant economic value. Indeed, the more energy intensive the material is, the more lucrative its recovery and reuse can be (for example, aluminum).[17]

Manufacturing operations can be deliberately designed to maximize waste and cost savings. The proper configuration of packaging lines, especially materials-handling systems, and the overall footprint of a line on the factory floor can lead to significant energy savings. Opportunities exist for packaging machinery manufacturers that find ways to help CPG manufacturers—their customers—lower costs related to packaging waste.[18]

Maximizing cost savings by eliminating waste requires a comprehensive approach. In particular, CPG producers' manufacturing operations—where installed equipment resides in the packaging supply chain—should be involved in this process. When major capital investment is planned, lean line design offers major opportunities to build cost savings and waste reduction into a packaging line.[19]

Major opportunities also exist in minimizing the time required to reconfigure lines for different packaging formats ("change parts"), which can reduce the resources wasted while a line is idle and not producing saleable goods. Automation, remote monitoring, modular design, and other innovations can all contribute in this context.[20]

The emphasis on reducing packaging wastes—and costs—through more sustainable manufacturing practices appears to have survived the recent economic downturn. A 2009 survey of 199 packaging professionals by *Food Engineering* magazine and Clear Seas Research found that, despite the recession, nearly half rated sustainability as extremely or very important for their company's packaging operations over the following two years. Fifty-seven percent reported that their employers had formal sustainability plans in place, and 71 percent identified reducing energy consumption and waste streams as major components of their sustainability plans.[21]

Consumer Attitudes

Consumer attitudes exert a strong influence on how retailers and CPG manufacturers throughout the world view packaging. In the United States, a broad and growing segment of consumers express a preference for products with more sustainable or "green" characteristics.[22] A recent study found that environmental considerations motivate 54 percent of American consumers to purchase products they perceive as green.[23] According to another study, purchases of environmentally friendly products by highly motivated consumers rose 4.1 percent from 2008 to 2009.[24]

As consumers adapt to difficult economic times, they appear to be shifting from brand-name products to those bearing retailers' generally lower-priced private labels or brands. Private label products seem to do especially well among environmentally-motivated consumers.[25] This shift is significant for sustainable packaging, because retailers have greater control over packaging for their own branded products. Major retailers, such as Wal-Mart Stores and Target, are known to pay particular attention to implementing more sustainable packaging for their private labels.

The degree to which environmental concerns motivate American consumers varies. Product taste and performance, convenience, and nutrition remain consumers' top priorities.[26] Although consumers place increasing value on protecting the environment, only a few appear to make sustainability the dominant consideration. More than a third of the consumers surveyed recently by the Grocery Manufacturers Association and Deloitte Consulting balanced sustainability with considerations such as price and convenience.[27] These attitudes are widespread, however, and distributed among consumers from all income levels, age cohorts, major ethnic groups, education ranges, and household sizes. Sustainability also represents an opportunity to attract customers who are not committed environmental shoppers, but who choose products they perceive as green when their other priorities are met.[28]

Consumer attitudes are even stronger in Europe and Asia. A recent study of consumer attitudes in 15 major economies found that more than 70 percent of those surveyed placed a high value on "living an ethical or sustainable lifestyle."[29] The study also found that consumers identified "less packaging" with a desire to remove clutter from their lives.

The connection between ethical living, sustainability, and packaging is not surprising. Packaging is frequently used to convey emotional, lifestyle, and values-oriented messages to consumers.[30] The interplay of values and packaging also highlights the challenges CPG manufacturers and their suppliers face in making packaging more sustainable. Implementing more sustainable packaging is more than a technical or engineering challenge.

Consumers' attitudes and preferences do not always match their purchasing behavior. Research suggests that consumers have doubts about some products' credibility as a more sustainable choice. Many also appear to be poorly informed about the sustainability choices that are available to them.[31]

Companies in the broader packaging supply chain are increasingly aware of shifting consumer attitudes toward sustainability. Sixty-three percent of industry respondents to the 2009 *Packaging Digest* and Sustainable Packaging Coalition survey on sustainability in packaging reported that customer requirements have the greatest influence on their pursuit of sustainability. Thirty-seven percent of more than 1,000 respondents also identified consumer requirements as a major driver for sustainability.[32]

The Role of Retailers

Major retailers in North America and elsewhere play a central role in creating demand for more sustainable packaging. Wal-Mart Stores, Target, Tesco PLC and numerous others are determined to benefit economically by reducing costs associated with environmental waste. They are adopting supply-chain management practices to achieve their environmental goals, including more sustainable packaging. Increasingly, these retailers require CPG manufacturers to package their products more sustainably. Given the purchasing power and global reach of the largest retailers, CPGs and their suppliers have little choice but to respond.

Wal-Mart Stores has taken an especially strong position in demanding that its supply chain address sustainability, taking on what a leading industry observer describes as the role of "de-facto regulator" of the packaging market. In September 2006, Wal-Mart announced that it intended to reduce packaging

in its supply chain by 5 percent by 2013.[33] Shortly thereafter, it introduced its Sustainable Packaging Scorecard (or the "Wal-Mart Scorecard"), which has become an important tool for Wal-Mart in evaluating the sustainability of its suppliers' packaging.

Wal-Mart asserts that by meeting its 2013 packaging reduction goal, the company's U.S. operations can avoid the emission of 667,000 metric tons of CO_2 and save 66.7 million gallons of diesel fuel.[34] *Supply Chain Management Review* estimates that those goals, if achieved, will result in cost savings for Wal-Mart of $3.4 billion. The company's declared long-term packaging sustainability goal is to become "packaging neutral" by 2025. Being packaging neutral means that "all packaging recovered or recycled at [Wal-Mart and Sam's Club] stores will be equal to the amount of packaging used by the products on their shelves."[35]

Pushing responsibility for reducing packaging onto its suppliers, by means of its Scorecard, indicates Wal-Mart's emphasis on extracting waste from the supply chain. As Tyler Elm, Wal-Mart vice president and senior director of business sustainability, told *Supply Chain Management Review*,

> We recognized early on that we had to look at the entire value chain. If we had focused on just our own operations, we would have limited ourselves to 10 percent of our effect on the environment and eliminated 90 percent of the opportunity that's out there.[36]

Wal-Mart uses a variety of techniques, in addition to its Scorecard, to implement its supply-chain strategy. Thirteen Sustainable Value Networks (SVNs), made up chiefly of outside experts—from Wal-Mart suppliers, academia, government agencies, and non-profit organizations—are used to generate innovative ideas. One of these SVNs is devoted specifically to packaging.[37] The company has begun holding an annual sustainable packaging exposition to introduce suppliers to new packaging technologies, including packaging machinery. Wal-Mart works intensively with suppliers of its private brands to reduce packaging consumption as well.[38]

Target, the second-largest retailer in the United States, is pursuing its own initiatives to reduce the volume and environmental impact of its packaging. Target puts particular emphasis on working with suppliers of its private label, non-consumable products, such as garments and housewares. According

WasteWise (*www.epa.gov/wastewise*) is an EPA-led public-private partnership program designed to help companies reduce and recycle municipal solid waste such as packaging, paper and corrugated containers, and selected industrial wastes such as batteries, oil filters, and non-hazardous inks and sludges. The program offers technical assistance, online tools, and public recognition for company efforts that reduce waste and lower costs while helping local communities.

to Dana Luthy, Target's director of packaging, those products offer greater flexibility for Target to design its own packaging in collaboration with "its certified packaging suppliers."[39]

Food and beverage packaging is more challenging, Luthy contends, because manufacturing and packaging are generally more automated processes over which individual retailers have less control. Nevertheless, the company reports successes, such as changing packaging for its in-house confectionary line to unbleached paperboard with recycled content. Target has also switched from using petroleum-derived polymer plastic packaging, for a number of bakery items, to materials made from polylactic acid, a plant-based polymer.[40]

Other global mass-market retailers, such as United Kingdom–based Tesco PLC, Marks & Spencer, and others, have set ambitious goals for achieving more sustainable packaging. Tesco's initial objective, for example, is to reduce packaging weight by 15 percent by 2010. Tesco acknowledges that weight alone may not be a sufficient measure of packaging sustainability and has announced that it is developing "a more comprehensive, long-term target" in 2010. In the interim, Tesco reports that it is working with more than 250 suppliers to reduce packaging for house-label and -branded products sold in its stores. According to Tesco, its efforts so far have resulted in savings of more than 80,000 metric tons of packaging, including a 19 percent reduction in packaging for its house-label dairy products and a 34 percent reduction in private label produce packaging.[41]

Regulation: Domestic and International
Many countries, especially in Europe, have adopted legislation, regulations, and standards that address packaging and packaging waste. The International Organization for Standardization (ISO) has begun to develop international packaging standards that are based on existing European standards. Although

these measures do not address packaging machinery directly, they exert a powerful influence on the market.

Currently, there is little regulation in the United States that addresses the sustainability of packaging. Federal regulations focus primarily on procurement by U.S. government agencies and environmental claims made by CPG manufacturers. Many states are developing legislation aimed at shifting the costs of recovering and recycling packaging waste to manufacturers and retailers and at reducing the burden on local governments.

Extended Producer Responsibility
Most international packaging legislation addressing sustainability and waste is based on the concept of extended producer responsibility (EPR). EPR forms the framework for European packaging legislation and is spreading in Asia. EPR, also known as product stewardship, holds that all parties involved in the various stages of a product's life cycle—including its packaging—take responsibility for mitigating its environmental impact. As a result, EPR programs generally include impacts connected with design, manufacturing, retailing, and consumption. Producers of packaging and other waste, which may be manufacturers, distributors, or retailers, are required to devise waste diversion or recovery systems for officially designated materials.

EPR-based programs frequently aim to encourage producers to take environmental considerations into account when designing new products and packaging.[42] "Design for the Environment" (DfE) for packaging can mean reducing the volume and weight of packaging, substituting less environmentally harmful materials, or incorporating new, more sustainable materials design. EPR programs do have a proven international track record of reducing the overall volume of packaging waste, although how much the programs result in DfE is open to question.[43]

EPA's **Product Stewardship Hub** (*www.epa.gov/epawaste/partnerships/stewardship/index.htm*) is an online informational resource designed to inform manufacturers, retailers, governments, and local communities on what product stewardship entails, product-service systems that foster better product stewardship by companies, and tips for how key industry sectors can reduce waste and manage products at their end of life.

U.S. Federal and State Regulations

The United States does not have comprehensive federal legislation addressing packaging and sustainability. The most relevant federal regulations are the Comprehensive Procurement Guidelines, which promote the use of materials recovered from solid waste by executive branch agencies of the U.S. government and by state and municipal government entities using funds appropriated by the U.S. Congress for procurement.[44]

The guidelines are authorized by the Resource Conservation and Recovery Act (RCRA) and by Executive Order 13423, and the U.S. Environmental Protection Agency (EPA) administers them. The RCRA requires EPA to designate products that can be made with recovered materials and to recommend practices for buying the products. Affected agencies are required to purchase designated products "with the highest recovered material content level practicable." The U.S. Department of Agriculture operates a similar program, BioPreferred, which designates bio-based products for use by federal agencies and their contractors in a "preferred procurement program."

Several U.S. states, including California, Maine, Minnesota, Vermont, and others, are developing legislation for packaging based on EPR. These efforts are driven by a desire to shift the cost burden for recycling away from municipal governments and state government agencies, toward manufacturers and other supply chain participants.

A number of U.S. states already operate EPR-based programs for toxic and other problem wastes. In response, a number of industries have established national recovery programs for materials associated with their products. Some of these private-sector recovery programs include the Rechargeable Battery Recycling Corporation, the Thermostat Recycling Corporation, and the Carpet America Recovery Effort.[45]

European Union

European governments began to address the management of packaging waste in the early 1990s. In 1994, the European Union adopted the European Parliament and Council Directive 94/62/EC, on Packaging and Packaging Waste ("the Packaging Directive"). While the Packaging Directive has a number of goals, source reduction is its principal objective. It also sets out ambitious targets for the recycling or reuse of packaging in EU member countries. Subsequent amendments establish definitions for packaging and set targets for the ten new member countries that joined the EU in 2004.

The Packaging Directive has led to the establishment of national recovery systems for discarded packaging in all 27 countries of the European Union. Several neighboring states have established such systems, as well.[46] The details of these national programs vary widely. In general, however, they mandate that manufacturers, retailers, and service businesses take back and recycle any packaging waste they generate. Retail, secondary, and transport packaging are all included under the directive.[47] Affected companies can perform these functions themselves or contract with a recognized service provider.[48]

International Standards

Another consequence of the Packaging Directive was the adoption in 2000, with subsequent revisions, of a series of European standards for packaging. These six standards, EN 13427 through EN 13432, address a range of requirements, including source reduction, reuse, recycling, energy recovery, and composting and biodegradation. These standards are intended to conform to the essential requirements specified in Annex II of the Packaging and Packaging Waste Directive.[49] In particular, requirement 1, which is "specific to the manufacturing and composition of packaging," stipulates that "packaging shall be so manufactured that the packaging volume and weight be limited to the minimum adequate amount to maintain the necessary level of safety, hygiene and acceptance for the packed product and for the consumer."[50]

The ISO has begun developing international packaging standards. The ISO's Technical Committee 122/Subcommittee 4 (TC122/SC4) on packaging and the environment, which met for the first time in December 2009 in Stockholm, will develop the standards based on the six European EN standards and on guidelines proposed by some Asian countries.[51] This effort follows calls from major CPG manufacturers and packaging industry organizations, such as the European Organization for Packaging and the Environment (EUROPEN), for consistent international packaging definitions, metrics, and standards.[52]

The ISO expects the packaging standards to be approved and published in 2012. U.S. involvement in developing them will be coordinated by a technical advisory group (TAG) accredited by the American National Standards Institute (ANSI). Interested U.S. stakeholders are encouraged to participate in the U.S. TAG, which is being administrated by the Material Handling Industry of America (MHIA), an ANSI member and accredited standards developer.

III. Sustainability as Competitive Advantage

Sustainability offers the greatest competitive advantage to packaging machinery manufacturers that focus on source reduction, specifically on technologies that reduce their customers' consumption of packaging materials. Because materials represent most of the cost of packaging, they offer a viable target for machinery manufacturers seeking a competitive edge. A common concern for companies considering a capital equipment investment to attain greater sustainability in packaging is that new machinery must also provide an appropriate return on investment. In addition to the cost savings from using less material, other benefits include reduced energy usage; lower greenhouse gas (GHG) emissions; and, in some cases, less water usage for producing the packaging materials.

Packaging machinery OEMs that pursue business strategies built around sustainability are responding to opportunity, rather than to demand (see section V, "Challenges to Implementing Sustainability"). Such companies leverage their positions as technology suppliers to identify and find ways to meet their customers' needs. "As a packaging machinery manufacturer, we are in a unique position to offer creative solutions," said Dale Andersen, president of Delkor Systems. "We hear the problems and challenges of our customers ... and then it is up to us to deliver real-world machinery solutions that work. Sustainability is causing a lot of companies to reconsider how they package and, as a result, this has been a great opportunity for Delkor."[53]

Packaging machinery OEMs profiled in this study (see section IV, "Manufacturer Case Studies") do a growing business in both domestic and international markets. Delkor Systems has a strong position supplying U.S. and Canadian dairy processors and also competes successfully in Mexico and South America

against more established European manufacturers. Hartness International operates in more than 100 countries on six continents. The Nordson Corporation sells equipment for dispensing adhesives and coatings around the world. A focus on reducing materials consumption enables these and other U.S. manufacturers to compete successfully on the basis of sustainability.

Packaging Materials as a Focus for Sustainability

From a business standpoint, a focus on packaging materials targets most of the value in the packaging supply chain. According to a conservative estimate, packaging materials account for roughly $111 billion of a roughly $125 billion annual market for packaging goods and services in the United States.[54] In contrast, sales of primary and secondary packaging machinery represent less than 5 percent of the total. Market participants interviewed for this study also believe that the expense of operating packaging machinery is minimal in relation to the cost of material, although such expenses have not yet been well documented.

Packaging materials offer considerable scope for innovation. Opportunities exist across a broad range of materials and formats, from corrugated board and plastic films to inks, adhesives, and line lubricants. These opportunities are not limited to large companies. Indeed, firms with less than $1 billion in annual sales—in some cases, considerably less—generate some of the most innovative sustainable packaging solutions.

Source reduction directed at packaging materials also aligns equipment manufacturers with the demand for sustainability in the rest of the packaging supply chain. From a market perspective, it means being responsive to major U.S. and European retailers' demand for more sustainable packaging. Cutting the use of packaging and eliminating packaging waste also reflect the objectives of the extensive regulatory measures found in Europe and elsewhere.

Automation

Automating packaging machinery and packaging lines offers a number of sustainability benefits. From an economic standpoint, automation enables CPG and contract packagers to achieve the throughput volume and consistency they need to be competitive.[55] Advanced monitoring and control capabilities enable them to maintain higher "up-time" for their packaging lines and to reduce packaging waste, energy consumption, and rejected packages. Ergonomic design enhances worker safety and reduces repetitive motion injuries and other hazards.

Advanced automation systems collect and apply real-time data about the performance of individual machines and entire packaging lines. Programmable control devices, feedback circuits, and servo drives can adjust performance without interrupting packaging operations.[56] As a result, operational problems can be headed off, and the waste of packaging materials, energy, and labor from equipment and line malfunction can be reduced.

Automation can also reduce time and costs associated with change parts. Servo technology enables some packaging machinery OEMs to offer change-over times of as a little as 5 to 15 minutes. Some industry observers believe that five-minute change parts may become a standard end-user expectation in the not-too-distant future.[57]

Automation can also enable packaging equipment and lines to adapt more quickly and efficiently to lighter-weight packaging materials and more variable material characteristics resulting from greater recycled content. "Material variability is one of the issues packaging machinery OEMs like to talk about," said John Kowal, market development manager for B&R Industrial Automation, a manufacturer of automated control systems for packaging and other capital equipment. "The more you build this capability into machines, the more you 'future proof' it against new materials."

The challenges posed by material variability can be seen in the case of corrugated board, which is the most widely used of all packaging materials. Recycled paper fiber can be used very effectively to produce corrugated board and other packaging papers. That said, fiber is degraded every time it is recycled, and fiber length is reduced, which results in paper with less strength than that produced with entirely "virgin" fiber.[58]

Recycled corrugated board "is much more variable than non-recycled material, especially at higher [machinery operating] speeds," said William Chu, general manager of Wexxar Packaging, which manufactures automated case formers and other end-of-line packaging machinery. The age, size, shape, and design of the corrugated material all affect a machine's operating characteristics. Recycled content can worsen the imperfections present in any corrugated board, causing additional warp, dimensional variations, and increasing fragility. "In less demanding environments, greater variances in corrugated can be addressed manually, but at higher speeds, these corrugated variances can impact the time it takes to open cases and can impact case flow across an entire production line, especially over long periods of operation," said Chu.[59]

The sustainability benefits of automation can be augmented with other technologies. Sensors and timers, for example, can be used to power down energy-intensive subsystems, such as the heated glue pots used in hot-melt adhesive dispensing equipment. Applying modern information and telecommunications technology makes possible remote, real-time reporting of an automated packaging line's operational status to a personal digital assistant or desktop computer.

Flexibility has become a major consideration too for the design of automated packaging systems. Major retailers have become very specific in recent years about the types of packaging they want from their vendors. Sizes, shapes, closures, and other features can vary significantly for identical products across major retail chains. Secondary packaging can vary widely as well. According to Andersen of Delkor Systems, "major merchandisers are now getting involved in selecting the style of shipping package that goes to market." Possible alternatives might include a standard tray, a standard case, a retail-ready display tray, or a pad-shrink shipper.

Accommodating flexibility can be a significant design challenge. Many older packaging machines were designed for only one type of packaging operation, such as forming a standard corrugated cardboard case. Newer machines can be designed

Mechatronics has been described by one of the field's leading journals as "the synergistic combination of precision mechanical engineering, electronic control and systems thinking in the design of products and manufacturing processes. It relates to the design of systems, devices, and products aimed at achieving an optimal balance between basic mechanical structure and its overall control."[a] Purdue University Calumet (http://webs.calumet.purdue.edu/technology) is the first U.S. institution to offer a degree in mechatronics engineering technology, although numerous others have courses, laboratories, and ongoing projects.

a. *Mechatronics, The Journal of the International Federation of Automatic Control.*

for greater flexibility, for example, with modular components that can be switched out quickly and easily to accommodate different packaging formats. More flexibility means more design complexity, however, which can slow down change parts operations.

Nevertheless, design innovation involving automation and flexibility offer OEMs and their customers important opportunities to add value and to reduce costs. Packaging functions previously considered stand-alone can now be combined into a single solution. Modular design can also accommodate retrofitting machinery with new capabilities as they are developed. As a result, OEMs and packagers can implement incremental upgrades to their packaging technology without replacing an entire machine.[60]

For packaging machinery OEMs, automation requires both in-house engineering capability and collaboration with manufacturers of process control technology. Successful automation also requires integrating mechanical and electrical technologies, which can be challenging for companies with small engineering departments. The development of a new discipline, mechatronics, which includes elements of mechanical and electrical engineering and computer science, is intended to address the needs of packaging machinery and other capital goods industries. The United States' first mechatronics engineering program has been established, with the support of numerous packaging machinery manufacturers, at Purdue University–Calumet's School of Technology in Hammond, Indiana.

Ancillary Products

Reengineering the application of ancillary products, such as adhesives, inks, and line lubricants, offers additional sustainability opportunities for packaging machinery manufacturers. Petroleum products—especially ethylene vinyl acetate (EVA) and metallocene resins—and water are two of the principal ingredients of most commercial adhesives. In absolute terms, the expense of these inputs may be modest relative to materials, such as paper, plastics, or metal. Nevertheless, the raw material costs and their environmental impact can still be significant.

A variety of process energy costs are associated with ancillary products. Hot-melt adhesive systems require electricity to maintain adhesives at optimal application temperatures. The most widely used packaging adhesives, which are based primarily on EVA, must be kept between 350°F and 375°F. Curing conventional applied ceramic labeling (ACL) for glass bottles can also be a major energy cost, because the inks must be baked onto the glass substrate at temperatures of 1,000°F or more for as long as two hours.[61]

Sustainability solutions for adhesives include reducing the amount of adhesives used for a given application through the redesign of application equipment or techniques and through automation. Some of these techniques include reducing the diameter of openings on adhesive dispensers and developing alternative adhesive dispensing techniques. Ultraviolet curable adhesives, especially for use with blister packaging and inks, also offer considerable energy savings.

A focus on ancillary products can also lead to significant conservation of water. Substituting dry lubricants for the large quantities of soapy water conventionally used as bottling line lubricant is another sustainability opportunity that can significantly reduce customers' costs.

Packaging Systems

Redesigning packaging systems is another approach that innovative packaging machinery OEMs are taking. In the United States and elsewhere, OEMs have leapfrogged over CPG manufacturers to design new package types and the machinery on which to run them.[62] This approach leverages an OEM's familiarity with packaging materials and its own customers to develop alternatives to conventional systems. Successful redesign can lead to significant costs savings for customers through reduced material consumption and transportation costs. New types of packaging systems can also open up new business opportunities for the OEM, such as helping customers procure optimal materials for use with the new packaging system.

Redesigned packaging systems identified in this study have some common elements beyond the savings realized through source reduction. Because less packaging material must be removed at the point of sale, these systems enable a quicker, easier transition from transportation and storage to retail. The plastic films and corrugated board of which they are composed can easily carry artwork, which supports a CPG manufacturer's efforts to build brand recognition. Alternatively, transparent films can make the product itself visible to consumers.

OEM Manufacturing Processes

OEMs' own manufacturing practices can also make a contribution to the overall sustainability of their businesses. PMMI encourages its members to consider implementing "lean" manufacturing concepts. Many reportedly have worked with the Manufacturing Extension Partnership, a program of the National Institute of Standards and Technology, to do this. PMMI staff estimate that as many as 40 percent of their member companies have continuous improvement programs.

Remanufacturing or, at least, refurbishing of packaging machinery also appears to be a common OEM practice. However, the long service lives of most packaging machinery and the pace of technological change in recent years complicates remanufacturing. With some machinery in use for 30 years or more, returning an older unit to its original condition may not make sense from a commercial standpoint. Remanufacturing a piece of equipment originally built around pneumatic or even simple mechanical technologies, for example, may require replacing everything but the unit's basic metal frame.

Best Practices

The packaging machinery manufacturers profiled in this study follow a number of business practices that support their focus on sustainability and innovation. The sustainability of their products is measured and documented using well-established methodologies, such as Total Cost of Ownership (TCO), Life Cycle Assessment (LCA), or Overall Equipment Effectiveness (OEE). Their product literature frequently makes reference to the SPC definition of sustainable packaging, which is a widespread, if unofficial, point of reference for sustainability throughout the packaging supply chain. Close relationships with key technology vendors are also common among these companies.

While most packaging machinery OEMs focus—understandably—on source reduction, they also have many opportunities to reduce costs and environmental wastes by making machinery more efficient. Reducing energy consumption, in particular, can be addressed in several ways: by replacing compressed air systems with servos and other electromechanical devices and by using better circuitry designs, among others (see Appendix 2 for more information). Installing electric power monitors also provides end users with a means to track and control energy consumption.

The bottom line of all these different approaches for packaging machinery OEMs is increasing sustainability by cutting their customers' operating costs. Reducing the consumption of packaging and ancillary materials has a variety of environmental benefits, but it also directly addresses sustainability's economic dimension. "At the end of the day, we all want to save [natural] gas, water, energy," said an engineer from a major U.S. CPG.

IV. Manufacturer Case Studies

A principal objective of this study is to help U.S. packaging machinery OEMs to be more competitive. Real-world examples are essential to successful communication about sustainability. Therefore, case studies of several packaging machinery OEMs that have developed significant sustainability programs have been included.

Companies in the following case studies have annual sales ranging from $1.5 million to nearly $900 million. These companies also represent a variety of business models: suppliers of integrated packaging lines, smaller machinery manufacturers with a tight focus on a specific segment, and a larger manufacturer of highly specialized products.

It should also be noted that companies, products, and services are cited in the case studies for illustrative purposes only—and not as recommendations or endorsements. Packaging machinery end users looking for specific packaging solutions should conduct their own due diligence, based on the full range of relevant vendors, to determine which suppliers, products, and services best meet their needs.

Delkor Systems, Inc.

Delkor Systems, Inc., of Minneapolis, Minnesota, manufactures automated end-of-line, secondary packaging equipment. The company's products include case and tray packers; top-load carton formers, loaders, and closers; shrink bundlers; and robotic systems. A small business, Delkor has 100 employees and approximately $40 million in annual sales.[63]

Delkor's management began investing in the late 1990s in the development of new packaging technologies to reduce the use of corrugated board in packaging. According to Delkor president Dale Andersen, the management asked themselves, "How can we change the package itself?" They realized that there was a "tremendous opportunity in reducing the corrugated that goes into packaging." Since then, Delkor has been awarded seven U.S. patents for new shipping package concepts that specifically provide greater efficiency in using corrugated board to transport product to market. The most successful new concept was a pad-shrink packaging system, Delkor's Spot-Pak® package, which is an alternative to the traditional corrugated box or regular slotted container (RSC) case.[64] This system is now widely used for consumer-packaged food products in a variety of package formats for plastic cups and bottles, paperboard containers, and other products.

As an alternative to the RSC—what a layperson might describe as a "cardboard box"— Spot-Pak® uses a temporary bonding adhesive to stabilize containers positioned on a flat corrugated pad. Containers can be stacked several layers high, according to the end user's requirements, with a corrugated pad between each layer. The final assembly is then shrink-wrapped in polyethylene film into a single bundle for shipping.[65]

To document the environmental impact of Spot-Pak®, Delkor commissioned Allied Development Corp., a specialized consulting firm, to conduct an LCA. According to the results of that study, which was made public by Delkor and its customer, Smart Balance, use of the company's pad-shrink technology could reduce the amount of packaging waste (by weight) to be recycled or deposited in a landfill by 82 percent, compared with standard corrugated RSCs. Reduced raw material input and material handling, in turn, could cut the use of process energy by 62 percent, largely through reduced transportation costs for polyethylene film compared with corrugated board. GHG emissions were calculated to be 55 percent lower because of reduced energy consumption during transportation and material processing and from reduced use of raw materials. Finally, energy consumption from transporting packaged goods to the point of sale was reduced by 11 percent because of greater product density in shipping.[66]

Andersen credits lower costs and rapid return on investment (ROI) with helping Delkor compete successfully in both domestic and international markets against other manufacturers, especially European companies with a strong historic presence in the Western Hemisphere. "The big benefit for our customers is we're going to achieve at least a 50 percent reduction in packaging," he said, as well as 8 to 12 percent higher product density on shipping pallets. Because of the lower costs that it offers— which Delkor asserts may be as high as $500,000 per installed system per year—Spot-Pak® also provides an ROI of as little as one to two years.[67]

According to Delkor, there will be nearly 200 Spot-Pak® systems installed in the United States, Canada, and Mexico by the end of 2010. The company reports that its equipment is used to package as much as 50 percent of the cottage cheese, sour cream, and yogurt products in North America. Major customers include the dairy manufacturing division of Safeway Stores and Smart Balance, Inc., which manufactures a national brand of buttery spread. The company has supplied a number of packaging lines to Mexico's largest dairy company, Grupo LaLa.[68]

Delkor also contends that its success with Spot-Pak® has led to other business opportunities. It recently introduced Tray-Pak®, a bundled shipper similar to Spot-Pak®, which converts into retail display trays once the film shrink wrap is removed. Although the company is not a converter, it can provide customers with procurement services for materials (corrugated pads and trays, polyethylene film and specialized adhesives) that work most effectively on its equipment. In addition to its full line of carton forming, loading, and closing equipment, Delkor manufactures robotic packaging solutions based on FANUC LTD (Japan) and KUKA Robotics (Germany) technologies.

Hartness International

Hartness International, an ITW Company, is a Greenville, South Carolina, manufacturer of secondary packaging machinery that has positioned itself aggressively in recent years as a provider of complete packaging solutions. Hartness's services include line automation and integration, project management, development, design, and installation. Acquired in November 2009 by ITW, Inc., Hartness employs 500 people and has annual sales of approximately $100 million.[69]

"Hartness is the most proactive in this space [sustainability] of any of the companies in the industry," said veteran packaging industry analyst and consultant Ben Miyares. Using what Miyares describes as "a pretty holistic approach" to sustainability, the company's management draws on its established automation and integration businesses to help its customers reduce their consumption of packaging materials and energy. The company offers robotic and other automated materials handling solutions designed to reduce damage to today's increasingly fragile, "light-weighted" packaging. Hartness also uses automation to provide greater changeover flexibility for customers running multiple formats on their packaging lines.[70]

Hartness develops new packaging lines with smaller footprints, designs based on Lean Enterprise concepts, and systems integration.[71] Vice president of operations Sean Hartness observed that the company's line integration and sustainability businesses developed in tandem. "As we became an integrator...to be competitive, we designed a line that fits into 20 to 30 percent less space, allowing us to put more production lines in smaller spaces."[72]

Hartness's director of global market development and emerging businesses, Scott Smith, contended that the company's lean packaging line designs have won business "not just because they meet someone's 'sustainability' criteria, but because they make good business sense and eliminate waste." One of their most successful beverage line installations was done using 30 percent less space than the industry standard, he reported. As a result, the line required 30 fewer motors, resulting in a 40 percent drop in utility costs. "We won," Smith continued, "not because we necessarily were more 'sustainable,' but because we had the best solution for our client's business. If you have both components, you are usually tough to beat."[73]

Specialized information technology applications augment automation and integration. Global Messenger is a real-time service that transfers data on a packaging line's operations to personal digital assistants and desktop computers for remote monitoring. The service is attached to the line's electrical control system, with software resident on the Internet; no external peripherals are required. Another service, FlashBack, is a high-speed video troubleshooting system that analyzes the movement of manufacturing equipment and processes and identifies problems as they develop.[74] The tools are aimed at reducing unscheduled service travel, as well as the costs and environmental impacts associated with travel.[75]

Another Hartness service, Life Cycle Sustainment, works with customers to extend the life and improve the capabilities of existing, installed machinery. Life Cycle Sustainment leverages a variety of services and technologies, with the objective of reducing waste and costs associated with older equipment.[76]

Hartness has also focused a great deal of attention on the sustainability of ancillary products. Dry Conveyor Lubricant (DCL) is a "non-aqueous" Teflon-based conveyor lubricant that is designed to eliminate the use of water-lubricating bottle, jar, and canning lines. According to Hartness, the mineral oil used as a carrier is "food grade, medicinal white oil" commonly used in medical and pharmaceutical applications. DCL contains no water and may save up to 250,000 gallons of water annually.[77]

For nearly five years, Hartness has been developing another line of ancillary products, its Uvaclear™ inks. These products are designed to reduce customers' energy consumption and GHG emissions by applying inks used on glass bottles that can be cured with ultraviolet light, because UV light consumes significantly less water and energy than conventional ceramic labeling. The inks are intended as an alternative to conventional ACL, which must be baked on at high temperatures for an hour or more and often contain heavy metals in their pigments.[78]

Hartness's sustainability activities also include developing a packaging system, Grab Pack, that uses a "preformed web of printed shrinkable polyethylene film" to bundle grouped containers for transport, storage, and display. The system reportedly uses up to 30 percent less film than conventional alternatives. Grab Pack bundles are placed directly on a pallet without the use of shipping cases or trays. The tightness and stability of the bundled package are designed to make it easier to handle in a variety of settings.[79]

Hartness also offers D-FOAM-R™, a patented technology that eliminates foam in bottling lines for carbonated soft drinks. D-FOAM-R™ uses ultrasound to break the surface tension of the bubbles that make up the foam. Eliminating the foam in carbonated soft drink production increases the speed and efficiency of bottling operations, eliminates the requirement to overfill bottles to offset foam loss, and reduces the rejection of underfilled bottles.

Hartness works closely with a number of materials suppliers to support its sustainable packaging products and services. In the case of Uvaclear™ inks, partners include INX International and Kammann Machines, Inc.[80] INX International, a leading global producer of inks, manufactures Uvaclear ink and provides technical support. Kamman Machines is a major manufacturer of printing equipment and supplies graphic screen-printing machinery, including high-speed equipment and decorating expertise. For Grab Pack, Hartness collaborates with Swedish film producer RKW Sweden, AB.[81]

Smith observed that the use of partnerships has been especially critical for Hartness for some of the company's new film-based products, Grab Pack in particular. "Our goal with this product is to reduce material and material waste," Smith said. "The Grab Pack is designed to utilize the minimum amount of film possible, while providing the maximum amount of pack integrity in a package that allows our CPG clients to differentiate themselves." Close cooperation with film manufacturers and converters was essential to the success of the Grab Pack package design and the development of the film characteristics and application technology.[82]

Hartness recognizes both OEE and TCO as valuable tools for evaluating the sustainability of packaging machinery. Smith cited Mean Time to Repair (MTTR), an OEE metric, as an example. MTTR describes the time required to repair a machine in the event of unscheduled downtime or malfunction. When trying to improve MTTR times, Smith said, Hartness looks at the whole process. "Our goal is to eliminate waste" and simplify processes. "We see these as very much a function of sustainability."[83]

Smith stressed, however, that total cost of ownership should be a determining factor in end-user purchasing decisions. "Aren't packaging machines that consume [fewer] parts more sustainable and more environmentally friendly than ones that consume significantly more?" Machines that discard fewer worn parts cost clients less to operate and ultimately reduce both the cost and environmental impact of operation.[84]

Hartness reports sales in more than 100 countries on six continents, with overseas offices in China, Germany, and Mexico. Key customers include TetraPak, Diageo, Coca-Cola, Heineken, Combibloc, Hennessy, SABMiller, Absolut, Kraft, Unilever, Chivas, Heinz, Tropicana, Nestlé, Procter & Gamble, Grand Chais de France, and Efes Pilsen.[85] More than 250 of Hartness's DYNAC conveyor systems are installed in Europe. Uvaclear™ inks and Grab Pack have been launched there as well. [86] The company's traditional case packaging machinery business still accounts for 10 percent of its annual sales.[87]

Nordson Corporation

Nordson Corporation of Westlake, Ohio, manufactures equipment for dispensing adhesives, sealants, coatings, and other industrial materials. A major component of the company's adhesives dispensing business is its packaging adhesives systems unit, which manufactures equipment for dispensing, testing, and inspecting adhesives for food and beverage containers.

Nordson, which employs nearly 3,700 people worldwide, reported total sales of $819 million in 2009. Adhesive dispensing systems, with sales of nearly $460 million in 2009, represented 56 percent of the company's sales. More than 71 percent of Nordson's business is outside the United States. The packaging adhesives division is in Duluth, Georgia.[88]

Nordson approaches sustainability with two broad objectives: to help its customers reduce their consumption of adhesives and to reduce the energy requirements of its adhesive dispensing systems. Many commonly used packaging adhesives are composed of petroleum products and are soluble in either oil or water. Reducing adhesive consumption can lower the cost of adhesives as a manufacturing input, as well as reduce the environmental impact of consuming petroleum and water. More efficient hot-melt application equipment can also significantly reduce the cost and environmental impact of these energy-intensive systems.[89] Nordson estimates, for example, that one barrel of oil and 7,560 gallons of water are conserved for every 315 pounds of adhesive not used.[90]

Nordson relies heavily on automation to improve its products' efficiency, to control costs, and to reduce environmental impacts. In particular, the company has developed automated filling systems for the adhesive tanks on its hot-melt dispensing systems. Automated filling is intended to maintain a steady flow of adhesive at more constant temperatures in the fill tank. A steady flow of adhesive, in turn, can eliminate temperature spikes in the adhesive, improve operational and energy efficiency, and reduce adhesive waste. The use of timers can keep hot-melt adhesives hot during packaging operations, but can power down the dispensing system at other times, which further conserves energy.[91]

Economic and environmental benefits can also result from greater precision in applying adhesives. According to Nordson, reducing the nozzle aperture by 1/64th of an inch can save 25 percent in adhesive volume, which can reduce adhesive consumption by as much as 2,400 pounds annually. Greater application precision can also reduce adhesive waste without compromising performance, especially on smaller packages.[92]

Nordson dispensing systems also enable end users to apply a wide range of adhesives. This ability is significant because newer hot-melt adhesives require lower temperatures than conventional materials, such as EVA-based resins. EVA adhesives must be heated to 350°F to 375°F, while newer formulations based on metallocene resins can be applied at 250°F.[93]

Lower application temperatures and the chemical characteristics of the metallocene resins require less process energy during dispensing and reduced downtime for cleaning and maintenance. Nordson produces dispensing equipment that enables packagers to use the newest adhesive materials available, although the company has not developed a specific line of equipment to dispense the products. "It's important that we are able to enable any material with the most efficient and value-added product on the market," said Rick Pallante, marketing development manager for the Nordson's packaging adhesives division.[94]

Other technological innovations offer sustainability benefits as well. For example, dispensing adhesive in a stitch pattern instead of a solid line reduces consumption without degrading performance. Applying adhesive mixed with an inert gas as a foam, after a considerable period of technological development, may reduce adhesive use and cost by as much as 40 percent.[95] Nordson reports that its

PatternJet™ dispensing system for labeling reduces adhesive consumption by 50 to 90 percent.[96]

Because it operates a global business, Nordson must conform to a variety of environmental laws and regulations. Among the most important are the EU's Waste Electrical and Electronic Equipment (WEEE) and Restriction of Hazardous Substances (RoHS) directives. Nordson views conformity with WEEE and RoHS as enhancing the competitiveness of its products in international markets. According to Pallante, "it is as important to meet these global standards as it is for our equipment to be UL [Underwriters Laboratories] and CE [European conformity] compliant." Meeting these criteria "is part of our product specifications and market strategies and allows us our premium product status."[97]

Nordson uses its own TCO-based calculator to show potential customers the cost benefits of Nordson products versus older equipment. Demonstrating cost savings, rather than sustainability, is the principal objective of the calculator.[98] Nordson is not aware of customers selecting their products because of their sustainability. Pallante believes customers choose a Nordson product or system "based on the value that it provides them economically and the efficiency it provides their operations."[99]

Partner Pak, Inc.

Partner Pak, Inc., a small business based in Huntington Beach, California, manufactures equipment for ultraviolet (UV) sealing of clamshell and blister packaging. Partner Pak systems are particularly well suited for use with consumer electronics, because they do not use radio-frequency technology that can damage sensitive electronics products.[100] Partner Pak's president and CEO Paul Appelbaum reports annual sales of approximately $1.5 million.[101]

Partner Pak's Simpl-Seal® packaging technology is designed as an alternative for radio-frequency, sonic, and heat-sealing processes. The company's equipment applies a thin layer of UV-sensitive adhesive between the top and bottom sections of the blister or clamshell package. A clear seal is created when the package is exposed to UV light in a curing tunnel. A variety of automated and manual application equipment is available.[102]

The principal competitive advantage offered by Partner Pak products is the significant energy savings that can be realized by using UV-curable adhesives rather than conventional alternatives. Partner Pak documents this advantage by measuring the comparative sustainability of both a particular package type and the machinery on which it is run. Specific measures include energy consumption, which is verified at the electric meter; heat produced as a by-product of energy consumption; and damage caused or avoided during sealing. Partner Pak reports that Simpl-Seal® packaging technology significantly reduces energy consumption, produces less heat, and eliminates damage caused by high energy sealing.

The DYMAX Corporation of Torrington, Connecticut, is Partner Pak's exclusive supplier of UV-curable sealants.[103] Special adhesive features include Blue to Clear, a material that turns clear after curing. When new plastic packaging materials are introduced, Partner Pak has access to Dymax's research and development laboratories for creating suitable adhesives.

The Shape Company and Superior Quartz Corp. supply Partner Pak with UV lamps and systems with the appropriate wavelength signatures. Epson Robots is Partner Pak's preferred strategic automation partner because of the quality of its products and the training service it offers internationally. Partner Pak can design a packaging system around a customer's choice of thermoformers, which convert sheet plastic into packaging.

Partner Pak reports that the use of Simpl-Seal® can reduce energy consumption by as much as 80 percent compared with conventional sealing technologies, since UV curing doesn't require heat. Other important savings derive from the fact that the Partner Pak technology requires no specialized tooling.[104]

Partner Pak president Paul Appelbaum cited Costco as an example of how sustainability has enabled the company to compete successfully. "Costco wanted to go green by eliminating PVC (polyvinyl chloride) and replacing it with maximally sustainable RPET (recycled polyethylene terephthalate)." Competitors' systems could seal PET derivatives, but not RPET, "the most eco-friendly plastic then available." According to Appelbaum, RPET was also attractive to Costco because it was "and continues to be less expensive than the PET derivatives that other methods could/can seal."

Partner Pak's low-energy UV technology also produced significant energy savings for Costco. "The energy savings actually reduced Costco's carbon footprint, when compared to other sealing methods," Appelbaum said. Manufacturers such as First Alert, Honeywell, and Gamestop also use Partner Pak systems to package consumer electronics products, because UV technology will not damage sensitive electronic components during packaging operations.[105]

Partner Pak has also invested in reengineering conventional packaging designs. An example is its Trapped Card™ packaging, an alternative to standard blister packaging. The blister design is a very common packaging format, in which a product is secured between a preformed plastic dome or bubble and a paperboard backing or card. The dome or bubble is attached to the card by stapling, heat-sealing, or gluing, or by other means. Alternatively, the blister folds over the packaged product in a clamshell fashion. The blister components can be formed from a variety of plastic resins. The card provides visibility when the package is displayed on a peg-board or other retail setting.[106]

According to Partner Pak, the Trapped Card™ format seals the plastic blister components directly to one another using the company's Simpl-Seal® technology. The blisters or clamshells are designed with a wide lip to trap the card—which has a cut-out that corresponds to the size and shape of the blister—between the sealed components, without applying adhesive to the card itself. A similar concept, the trapped blister, traps a flanged blister between two cards, one with a cut-out that allows it to fit around the blister and the other a solid card. In each design, the cards receive no adhesive or other chemical treatment, which makes them much easier to recycle.[107]

Partner Pak reports that its machinery and adhesives are CE certified, which is sufficient for sale in Europe and other foreign markets. At the moment, according to Appelbaum, the countries to which Partner Pak exports do not require WEEE or RoHS compliance for its products.

Pro Mach, Inc.

Pro Mach, Inc., of Cincinnati, Ohio, is one of the largest providers of integrated packaging solutions in North America. Pro Mach offers packaging machinery and packaging line integration for customers around the world that manufacture food, beverages, pharmaceuticals, electronics, and other consumer and industrial products. The company has 11 subsidiaries, including Wexxar Packaging. The subsidiaries are organized into three divisions: primary packaging, end-of-line packaging, and identification and tracking. Pro Mach employs 900 people in North America and Europe and reported revenues of $250 million in 2008.

Pro Mach's president and CEO, Mark W. Anderson, identified four methods his company uses to deliver more sustainable packaging solutions: collaboration, innovation, customization, and integration. Collaboration involves working closely with customers, converters, and other vendors "to implement a complex system solution." Innovation is developing a better idea and implementing it effectively. Customization is crafting a solution to solve a specific problem, and integration involves implementing an integrated machinery solution for a particular problem. Anderson stressed that introduction of a more sustainable package "often makes the packaging problem more complex" and said that all stakeholders need to be involved in executing an effective solution.[108]

Wexxar Packaging in Delta, British Columbia, manufactures case and tray forming and case sealing machinery. Wexxar's BEL brand features end-of-line corrugated box sealers, tapers, formers, and packing systems. Wexxar's machinery has been installed in nearly 40 countries worldwide. Applications include agricultural products, packaged foods, household products, pharmaceuticals, and electronics.[109] In addition to its Wexxar and BEL divisions, Wexxar acquired IPak Machinery Ltd., a leading designer and manufacturer of corrugated tray forming equipment, in June 2009.

In 2009, Wexxar announced that it was upgrading its WF30 case former with an electromechanical servo system to provide reliable automated case forming despite varied quality and types of corrugated board. The presence of recycled content in corrugated board affects a machine's performance and can lead to imperfections including warping, changing dimensions, and greater fragility.

The addition of a servo system and advanced control technology enables the WF30 to measure and react to the resistance corrugated board cases present as they are being opened. The system makes continual, automatic adjustments to the WF30's "Pin and Dome" case opening technology. This capability helps maintain operating speed regardless of the

---

U.S. Department of Commerce, International Trade Administration

quality or thickness of corrugated board or the size of cases being opened. The WF30 is also designed for toolless set-up, making adjustments for different case sizes faster and simpler.[110] The servo system ties into a control platform configured to allow remote monitoring. The WF30 incorporates Festo servo technology and an Allen-Bradley control platform.[111]

These innovations also enhance the WF30's energy efficiency. Improved mechanical design, replacement of pneumatic components with electric or servo-electric drives, and, where appropriate, the use of more energy-efficient pneumatic components all reduce energy consumption. Wexxar's general manager, William Chu, reported that together these improvements cut the use of compressed air by as much as 60 percent.[112]

Wexxar also attempts to reduce consumption of electricity by building a "sleep mode" into most of its machinery, Chu reported. Wexxar products are designed to shut down all their motors if there is no demand and to start automatically once a demand is made. This feature not only saves energy but also reduces wear and tear on the machinery itself.[113]

Another Pro Mach subsidiary, Fowler Products Company, LLC, has worked closely with major beverage manufacturers to reduce the volume of material required for capping disposable plastic bottles. Located in Athens, Georgia, Fowler manufactures machinery for bottle washing and capping, cap handling, and cap inspecting.

Light-weighting has dramatically reduced many bottles' neck heights and support rings. As a result, from a packaging standpoint, bottles increasingly behave more like bags than bottles. Reduced bottle weight presents multiple challenges for machinery. New molds must be made for the bottle blow molders; fillers, cappers, and cap-handling systems must be modified; and material handling must be adapted to accommodate decreased structural rigidity. For high-profile consumer products such as soft drinks and bottled water, Fowler involved third-party package design firms as well.[114]

Fowler worked with their CPG manufacturer customers and other vendors for over two years to implement new capping systems. According to Jack Aguero, Pro Mach's vice president for business development and marketing, this effort led to orders for 50 retrofitted capping heads in 2009. Customers reduced their use of plastic by one-third, which equates to a savings of 95 million pounds, worth more than $60 million.[115]

President and CEO Anderson stressed that for Pro Mach, success depends on involving all key component vendors early in the design process and recognizing the complexity that more sustainable solutions often entail.[116] Wexxar General Manager Chu agreed: "Our design and engineering team always includes representatives from our technology partners in the development of new systems and products. We rely on them for their specific areas of technical knowledge and we expect them to consider any unique requirements from our industry."[117]

V. Challenges to Implementing Sustainability

Despite the competitive success some packaging machinery OEMs are having with sustainability, a number of challenges stand in the way of a robust market for their products and services. The most fundamental challenge is that there is little or no discernable demand for packaging machinery with specific sustainability characteristics beyond conventional performance attributes such as reliability and safety. Procurement practices for packaging machinery have not changed to match the growing emphasis that CPG manufacturers and large retailers have placed on sustainability. Moreover, no definitions, standards, or certifications exist to help packaging machinery manufacturers or their customers identify more sustainable packaging machinery. A number of methodologies are used to measure the cost, effectiveness, and environmental impact of packaging machinery, but no standard analytical framework is applied throughout the industry.

Demand

There is a widespread perception in the packaging machinery industry that no demand exists for more sustainable packaging machinery. Several reasons can be cited for this apparent lack of demand, which has been well documented by PMMI.[118] Nevertheless, it is likely that OEMs will begin to encounter demand in the relatively near future as consumer product manufacturers and other packaging machinery end users increasingly seek to cut material use, energy consumption, GHG emissions, and other impacts of their manufacturing operations.

Demand—The Status Quo

Executives from packaging machinery OEMs, representatives of PMMI, and leading private-sector industry analysts all underscore the lack of demand for sustainability. Nordson's Rick Pallante remarked that "customers so far are not putting sustainability as the number one criteria for choosing a packaging machine." An executive at another OEM noted that "the [equipment] manufacturing community has not had the demand placed on it yet" in the way that CPG manufacturers have.[119]

A study by PMMI clearly documents that machinery characteristics identified with sustainability generally are low priorities for most CPG procurement teams. In its 2007 Packaging Machinery Purchasing Process Survey of 164 CPG purchasing managers, characteristics such as total cost of operation, ergonomics, and innovative design were identified as second-tier priorities. Energy footprint and machine footprint size were ranked as third-tier priorities. Respondents identified only two characteristics as top priorities: machine reliability and safety characteristics. Although reliability and safety contribute to a machine's overall sustainability, it is clear that other sustainability attributes do not carry much weight when purchasing decisions are being made.[120]

The apparent lack of demand for more sustainable packaging machinery illustrates another widely reported phenomenon: the discrepancy between many packaging machinery end users' corporate sustainability vision, as articulated by senior management, and that vision's application to business operations such as procurement or finance. "Very few companies," according to a study by Deloitte Consulting, have "undertaken the work of integrating sustainability across functions." As industry analyst Ben Miyares put it, "when it comes down to the transaction consideration of machine 'A' or machine 'B,' sustainability is not yet a deciding factor."[121]

Frequently, CPG manufacturers disperse responsibility for the various costs associated with a piece of machinery over its lifetime among different parts of their business. The procurement team or division may be responsible for the initial capital costs, manufacturing for operating expenses, human resources for training employees to operate and maintain it, and so on.[122] Most of them don't take a holistic approach," Miyares noted. Procurement decisions in particular, he said, tend to focus on delivering the specified packaging capacity for the lowest price.[123]

The Deloitte study contends that corporate vision also falls short in practice when management fails to integrate sustainability into a company's business cases and supplier agreements. "Since sustainability is about ongoing year-to-year resource consumption and emission," management should take a longer-term and more comprehensive view in evaluating investments in sustainability. In particular, greater consideration needs to be given to broadening the costs and time horizons considered, as well as the range of benefits that may be derived from greater sustainability.[124]

Another barrier to implementing corporate sustainability visions, argued SPC executive director Anne Johnson, is that financial managers often exercise effective control over procurement. In such cases, she said, very short-term bottom-line financial objectives outweigh sustainability and many other considerations. John Kowal of B&R Automation agreed, "U.S. CPGs tend to want the safe choice, not the most innovative" because of their very cautious approach to financial risk.[125] There is "a massive need," Johnson said, for cross-training engineering, financial, environmental, and other professionals.[126]

Demand—Why It's Likely to Change
Demand for sustainability in packaging machinery is likely to develop in the near future, the status quo notwithstanding. Packaging machinery end users must already document—and improve on—their environmental and cost performance to satisfy the requirements of a growing number of major retailers, regulators, stockholders, and financial institutions. These requirements will only grow with the passage of time. Reporting is likely to extend to CPG companies' entire manufacturing operations, including packaging and packaging machinery. "More and more sustainability will be required of packaging machinery manufacturers over time," said Pro Mach vice president Jack Aguero.[127]

Many countries already impose both reporting requirements and packaging fees as part of their EPR-based packaging waste regimes. The type of information required and the level of detail vary widely. Sometimes the fees can be reduced—for example, for packaging that contains recycled content—with the proper documentation (for example, in France). In the United States, packaging fees are likely to be a common feature of EPR legislation being considered in a number of states. Retailers such as Wal-Mart Stores, Safeway, Tesco, and others already require many of the same categories of data from vendors. As Victor Bell, president of Environmental Packaging International, said, "now is the time to be EPR-ready."[128]

Cutting energy use and GHG emissions plays a key role in the growing demand for data reporting and transparency. For CPG manufacturers doing business in the European Union, major Asian markets, and elsewhere, pressure to implement low-carbon manufacturing practices is strong and growing. Jordan Berkley, the Apriso Corporation's director of product manufacturing execution systems, believes CPG manufacturers in the United States will face similar pressures. Already, many are pushing to "correlate energy [use] to operations across functional silos" in their manufacturing operations.[129]

Automation technology can make reporting easier for packagers because it generates and uses real-time information to achieve the greatest operating efficiencies. This data can also be used to meet retailer and regulator demands. Metering food processing and packaging machinery's electricity consumption is not yet a common practice, but "you can do it strategically and start tying energy consumption to units of production," said Mike Steur, of Hixson, Inc., an engineering firm specializing in food and beverage manufacturing. "The more sophisticated companies are anticipating the change and laying the groundwork now for measuring and reporting their emissions."[130]

David Dixon, senior director—strategic accounts for the Food and Consumer Products Group of Burns & McDonnell, agreed. His firm, a leading engineering services provider, has teams conducting energy audits in as many as 50 manufacturing plants. "We're looking at every piece of equipment," including packaging machinery, Dixon said. "We have a lot of work installing additional metering." Dixon stressed that, more and more, packaging machinery OEMs "need to bring a sustainable solution to their customers, and it will involve every aspect of their products."[131]

Even if most OEMs do not yet encounter sustainability criteria in requests for proposals, the case studies in section III of this report demonstrate that packaging machinery end users are already receptive to machinery that offers cost savings through source reduction. Energy efficiency, automation, and the use of real-time data to optimize effectiveness are all features of those companies' packaging solutions. Partner Pak's Paul Appelbaum offers this perspective on demand for more sustainable packaging technology:

Every prospective customer and current customer is or has been in the "we do not care" group. However, when [they] learned that [our] product requires less energy consumption and the elimination of tooling, their "do not care" attitude changes to "Really? Can you prove it?"[132]

Definitions, Standards, and Certifications

Another sustainability challenge facing packaging machinery OEMs is the lack of relevant definitions, standards, or certifications. The definition of sustainable packaging developed by SPC is used more and more frequently throughout the supply chain (see Appendix 1). Delkor Systems and Nordson Corporation both refer to it when discussing sustainability.

Definitions

To date, there is no definition of sustainability for packaging machinery or other relevant capital equipment, as such. The SPC definition, however, can accommodate packaging machinery. Relevant criteria in the SPC definition refer to packaging that is

- Able to meet market criteria for performance and cost
- Sourced, manufactured, and transported using renewable energy
- Manufactured using clean production technologies and best practices
- Physically designed to optimize materials and energy

Meeting market criteria for performance and cost are essential for any packaging machinery OEM. Sustainability strategies described in section III of this study already focus on the development of clean production technologies and best practices for the benefit of CPG manufacturers and on physical design to optimize materials and energy. The use of renewable energy is no less a challenge—and a potential opportunity—for OEMs than for companies in other parts of the packaging supply chain.

Standards, and Certifications

In the absence of a definition for sustainability in packaging machinery, it should not be surprising that there are no specific U.S. standards or certifications either. In their absence, European legal requirements for packaging, electrical equipment, and hazardous waste recovery are shaping the global marketplace. As noted previously, European standards for packaging materials serve as the model for international standards being developed by ISO. Although conforming to these measures can be onerous, companies that do so successfully are more likely to internalize sustainability practices that can make them more competitive globally.

In addition to the EU Packaging Directive and related statutes, other European laws that push U.S. exporters to internalize sustainability in their business operations include the WEEE and RoHS directives. WEEE mandates the recycling and reuse of many types of consumer and industrial products. RoHS requires that safer alternatives be substituted for hazardous materials, especially heavy metals.[133] With reference to capital equipment, the WEEE directive includes equipment for spraying, spreading, and dispersing liquids or gases and for monitoring and control panels for industrial equipment.[134]

Complying with the WEEE and RoHS directives can be expensive and time-consuming, but U.S. companies that go to the trouble have the opportunity to internalize these requirements in their business practices. As a result, they typically are better positioned to compete in the United States as their U.S. customers integrate WEEE- and RoHS-like requirements into their domestic business practices. Nordson's Rick Pallante observed that "for Nordson to market its products globally, we must be able to meet criteria [such] as WEEE and RoHS.[135]

Measurement

In the absence of widely accepted metrics that address the sustainability of capital equipment and a standard for sustainability in packaging machinery, some OEMs are adapting at least one of several available measurement tools to document the sustainability characteristics of their products. OEMs profiled in this study used one of three tools: Total Cost of Ownership (TCO), Life Cycle Assessment (LCA), and Overall Equipment Effectiveness (OEE). To help U.S. manufacturers measure sustainability more effectively, the U.S. Department of Commerce's Sustainable Manufacturing Initiative team is working with the Organization for Economic Cooperation and Development (OECD) to develop a sustainable manufacturing metrics toolkit, which will consist of a core set of sustainability indicators to be freely accessed by U.S. companies across industries.

Total Cost of Ownership

TCO is a methodology used in many industries to capture the fully burdened, lifetime operating costs of any capital investment, including packaging machinery. Lower TCO can result in greater sustainability, but it often requires greater initial investment in equipment, training, and systems.[136]

PMMI describes TCO in terms of an iceberg, with the initial capital cost of purchasing a piece of equipment represented by the tip that is visible above the surface of the water. According to PMMI, the tip of the iceberg represents only about one-seventh of the total cost of ownership for packaging machinery. The balance incorporates "sustaining" costs, including direct labor, utilities, consumables, and reliability, and "maintenance" costs, such as spare parts, servicing, and decommissioning.[137]

Appendix 3 provides one such TCO framework, designed by Deloitte Consulting and Hartness International. This framework incorporates both traditional cost metrics and core sustainability metrics to reveal the operational and environmental costs of a capital equipment investment. The first section focuses on initial costs associated with a capital investment, while the second quantifies the long-term natural resource and maintenance costs.

Life Cycle Assessment

LCA is "a method for characterizing impacts associated with the sourcing, manufacture, distributions, use, and disposal of a product or product system."

LCA is an internationally recognized methodology documented in ISO standards 14040 and 14044. A far more involved measurement tool than TCO, LCA's focus is environmental impacts.

LCA offers a great deal of flexibility in that it can evaluate a variety of impacts, track material and energy flows, and identify environmental "hotspots" in a process. Because of this flexibility, however, an effective LCA must be well designed, and the goals and scope must be well planned. LCAs also require large amounts of data, with the quality of analysis depending to a great extent on the quality of data.[138] As a result, the cost of a robust LCA can be significant.

LCA is widely used in Europe and is becoming more common in the United States. The Consumer Goods Forum, an international body that brings together major U.S. and other consumer product manufacturers and retailers, is identifying and adapting "shared global industry metrics" for packaging based on LCA through its recently launched Global Packaging Project. PepsiCo uses an LCA-based methodology, PAS 2050, to develop "carbon footprints" for their products, including its Tropicana orange juice and Walkers potato chip brands. A number of environmental non-governmental organizations (NGOs) also base their sustainability calculators on this methodology, including the Environmental Defense Fund Paper Calculator, the World Wildlife Fund Footprint calculator, and others.[139]

Overall Equipment Effectiveness

A third and final methodology used by packaging machinery OEMs is OEE. OEE evaluates the effectiveness of a specific manufacturing operation, over time, based on several measurement criteria. This methodology is frequently used in implementing lean manufacturing practices.

OEE is calculated using three basic metrics: availability, performance, and quality. Availability represents the percentage of time the equipment or operation was running compared to the available time. Performance measures the running speed of the operation compared to its maximum capability, often called the rated speed. Quality describes the number of good items produced compared to the total number of items produced.[140]

The value of this methodology lies in the analysis and comparison of a machine's OEE in one period versus another. Understanding the complete OEE breakdown—according to the three metrics, across time—can reveal opportunities for improvement. With its origins in lean enterprise concepts, OEE can be an effective tool for identifying waste in packaging and other manufacturing operations.[141]

How to Choose the Right Method
Depending on what a firm wishes to communicate, TCO, LCA, and OEE are each helpful in different ways. TCO lends itself to analyzing the economic and environmental costs of a capital investment during its "use phase," that is, from the time of initial procurement to the equipment's end of life.[142]

LCA, because of its focus on total impacts, lends itself to conducting a full environmental analysis of a product. Through linking specific impacts to costs, it can help firms strategically reduce waste and design more resource-efficient products and services. Because U.S. manufacturers are increasingly expected to document the environmental impact of their products, LCA is likely to find much more widespread application in the United States in the future.

OEE, while much less nuanced than a TCO analysis or an LCA, can help a manufacturer identify natural resource waste in a selected manufacturing operation. When equipment performance and product quality are less than optimal, opportunities for waste minimization and sustainability enhancements typically exist.

Other Resources
Two other sustainability metrics resources cited by firms consulted for this study include the Global Reporting Initiative (GRI) and SPC Sustainable Packaging Indicators and Metrics Framework. The GRI offers sustainable development reporting guidelines to help companies track performance related to environment, labor, human rights, anticorruption, and other corporate citizenship issues. Because the guidelines have been, and continue to be, developed by numerous stakeholder groups—businesses, NGOs, international organizations, academia, and others—the GRI has been successful in attaining a high degree of credibility with many multinational firms and some smaller companies.[143, 144]

The SPC Sustainable Packaging Indicators and Metrics Framework provides a set of common indicators and metrics to help companies measure progress against the criteria laid out in the SPC Definition of Sustainable Packaging (see Appendix 1). Stand-alone modules are provided for each component of the criteria: material use, energy use, water use, material health, clean production and transport, cost and performance, community impact, and worker impact. The toolkit also provides a set of user guidelines meant to assist users in obtaining and deciphering their sustainability data.[145]

Tips for Packing Machinery OEMs

- Build a sustainability strategy around source reduction for your customers, including packaging materials, ancillary products, and energy and water.

- Seek ways to minimize—and document—your products' energy consumption in anticipation of likely future customer requirements.

- Use a recognized methodology to document your machinery's sustainability attributes, such as Total Cost of Ownership, Life Cycle Assessment, or Overall Equipment Effectiveness.

- Benchmark your products and services relative to the Sustainable Packaging Coalition's definition of sustainability, an informal but widely recognized packaging industry benchmark.

- Form strategic relationships, as appropriate, with converters or other suppliers of packaging, ancillary products, machinery, or automation.

- Design for sustainability, including some or all of the following:

  - Disassembly of machinery, to provide customer with greater line layout flexibility and to aid in end-of-life remanufacturing or recovery, including parts and components.

  - Reduction or elimination of empty container transport—for example, locating of bottle forming equipment closer to or in the same facility as filling lines to eliminate shipping empty bottles by truck.

  - Use of more efficient shapes—for example, square versus round for more efficient shipping and storage.

  - Use of bio-based polymer materials on your machinery.

Note: This list was prepared with the assistance of Todd Bukowski, Packaging & Technology Integrated Solutions, LLC.

VI. Conclusion

U.S. packaging machinery manufacturers can offer their customers more sustainable packaging technologies and be competitive in doing so. A number of OEMs are successfully selling machinery, services, and related products based on sustainability. These companies focus their efforts on enabling their customers to cut costs through source reduction. They offer goods and services that become part of a buyer's strategy to reduce consumption of packaging materials, ancillary products, and energy and water during manufacturing, rather than allow products to be expended as waste for possible recovery or disposal later.

Innovative OEMs' recognition of and response to opportunity drive successful business strategies based on sustainability. These manufacturers' knowledge of the challenges their customers face fosters the search for innovation. Their solutions are based on research and development of innovative machinery, materials, and package designs—often in collaboration with other manufacturers.

Packaging machinery OEMs that pursue sustainability are aligning themselves with the broad direction of the packaging market. The larger global packaging and CPG supply chains are becoming highly mobilized around sustainability. Major retailers recognize that source reduction can yield significant cost savings and are demanding it from their suppliers. Both CPG manufacturers and major retailers are managing their supply chains to reduce costs associated with environmental wastes, including packaging waste. Consumer attitudes and investor expectations that favor more environmentally friendly products and packaging also provide motivation to include sustainability in business planning.

Regulation, too, is a growing influence on the global marketplace for packaging technologies. EU directives, European national laws, and standards that address sustainability are increasingly shaping the international regulatory environment. Even in the United States, companies that have met European and other international requirements in their overseas markets are internalizing the resulting manufacturing and business practices in their domestic operations.

As a result, these forces are driving manufacturers of consumer products and other packaged goods to rigorously document environmental impacts and related costs. In the not-too-distant future, sustainability-related reporting is likely to extend to manufacturers' entire manufacturing operations, including packaging. The growth of legally mandated recovery regimes for packaging waste based on EPR will only reinforce this trend.

Despite the beginning of the transformation described above, sustainability has been slow to reach the packaging machinery industry. A few innovative OEMs have succeeded in identifying and taking advantage of opportunities to sell more sustainable packaging solutions, often to strategic customers that may themselves be early adopters of more sustainable manufacturing practices. However, the majority of CPG manufacturers do not appear to assign a high priority to sustainability when procuring packaging machinery.

It has been well documented that many end users of packaging machinery have not integrated their corporate sustainability visions into business operations, such as procurement of capital equipment. Moreover, no definition, certifications, or standards currently exist for sustainability in packaging machinery. Furthermore, there is a widespread, if poorly documented, belief that packaging machinery consumes too little energy to be significant.

Nevertheless, demand for greater sustainability is all around the packaging machinery industry. Upstream, converters face strong expectations for less expensive, more environmentally friendly materials. Downstream, CPG manufacturers face ever more demanding requirements to deliver packaging solutions that save retailers money and reduce waste. They all operate in a legal and regulatory environment that increasingly mandates sustainability. The recent recession has not diminished these ongoing developments.

There are, in fact, good reasons to believe that packaging machinery OEMs will eventually face demand for more sustainable solutions. The OEMs cited in this study already appear to be competitive in employing technologies and equipment that cut costs associated with packaging waste. Ever more rigorous efforts to measure and reduce product life-cycle costs will likely bring the expense of operating packaging machinery into greater focus in the near future. Therefore, it is not too soon for packaging machinery OEMs to prepare for the day when their customers begin to demand machinery, services, and other products that can deliver cost–effective, sustainable packaging solutions.

VII. For Further Consideration

This study draws heavily on the experience and perspectives of individuals and organizations from throughout the packaging supply chain. In addition to their views on the current state of the industry, many stakeholders also identified actions that might make it easier for packaging machinery OEMs to implement and commercialize sustainable packaging solutions. These actions include the following:

- *Greater recognition of the role of packaging machinery in the supply chain.* Recognize the central—and often highly innovative—role packaging machinery OEMs play as technology suppliers to the supply chain in areas such as source reduction, energy efficiency, and packaging systems design.

- *Harmonization of metrics.* Include representatives of the packaging machinery industry in international efforts to harmonize definitions, metrics, and standards for the packaging supply chain.

- *Cross-disciplinary training.* Develop and implement cross training of professionals from a variety of disciplines that are involved in making decisions related to sustainability and packaging, including packaging engineers and managers, packaging designers, manufacturing engineers and managers, procurement managers, and corporate finance managers.

- *Voluntary certification.* Develop a voluntary sustainability certification for packaging machinery. A well-designed certification could serve as a resource-efficiency benchmark and help raise the market profile of more sustainable packaging machinery.

- *Sustainable packaging machinery standard.* Consider whether development of a sustainable packaging machinery standard might serve the interests of OEMs and their customers as sustainability becomes more thoroughly integrated into their manufacturing operations.

U.S. Department of Commerce, International Trade Administration

Notes

1. The Competitiveness Institute, "On Competitiveness," The Competitiveness Institute, Barcelona, 2010.

2. World Packaging Organization, "Position Paper: Market Trends and Developments," World Packaging Organization, Stockholm, 2008, p. 1.

3. Jeremy Barker, Louis Besland, Daniel Mahler, and Otto Schultz, *"Green" Winners: The Performance of Sustainability-Focused Companies during the Financial Crisis* (Chicago, Illinois: A.T. Kearney, 2009).

4. U.S. Environmental Protection Agency, "Municipal Solid Waste Generation, Recycling, and Disposal in the United States: Facts and Figures for 2008," Environmental Protection Agency, Washington, DC, 2010.

5. Executive Office of the President, Office of Management and Budget, *North American Industry Classification System, 2007* (Springfield, VA: National Technical Information Service, 2007), p. 424.

6. International Trade Commission, *Harmonized Tariff Schedule of the United States 2009–Supplement 1 (Rev. 1)* (Washington, DC: U.S. International Trade Commission, 2009), chapter 84.

7. U.S. Census Bureau, *2007 Economic Census* (Washington, DC: U.S. Census Bureau, 2007).

8. Ibid.

9. Impact Marketing Consultants, Inc., "The Marketing Guide to the U.S. Packaging Industry," Impact Marketing Consultants, Inc., Manchester Center, VT, 2006, p. 41.

10. International Trade Commission, U.S. Department of Commerce.

11. Ibid.

12. Walter Soroka, *Fundamentals of Packaging Technology, Fourth Edition* (Naperville, IL: Institute of Packaging Professionals, 2009); *Fundamentals of Packaging Technology* (Naperville, IL: Institute of Packaging Professionals, 1994); and Walter Soroka, *Illustrated Glossary of Packaging Terminology* (Naperville, IL: Institute of Packaging Professionals, 2008).

13. Laura Bix, Nora Rifon, Hugh Lockhart, and Javier de la Fuente, *The Packaging Matrix: Linking Package Design Criteria to the Marketing Mix* (East Lansing, MI: Michigan State University, 2003).

14. Impact Market Consultants, Inc., "The Marketing Guide to the U.S. Packaging Industry: 2006–2008 Edition," pp. 5–6.

15. Ibid., pp. 5–6.

16. Deloitte Consulting, Grocery Manufacturers Institute, and Packaging Machinery Manufacturers Institute, "Sustainability from Boardroom to Breakroom," Deloitte Consulting, Washington, DC, 2008, p. 18.

17. Packaging and Technology Integrated Solutions, "Understanding and Executing Sustainability Initiatives and Sustainable Packaging Programs," Packaging Strategies, Kalamazoo, MI, 2007, pp. 66, 113–115.

18. Packaging Machinery Manufacturers Institute, "2007 Packaging Machinery Purchasing Process Survey," Packaging Machinery Manufacturers Institute, Arlington, VA, 2007, p. 27.

19. Deloitte Consulting, Grocery Manufacturers Institute, and Packaging Machinery Manufacturers Institute, "Sustainability from Boardroom to Breakroom," pp. 11–15.

20. Ibid., pp. 8, 11–14, 26.

21. Kevin T. Higgins, "Flexibility on a Budget," in *Food Engineering* (Troy, MI: BNP Media, 2009), pp. 79–85.

22. Grocery Manufacturers Association and Deloitte Consulting, "Finding the Green in Today's Shopper: Sustainability Trends and New Shopper Insights," Grocery Manufacturers Association and Deloitte Consulting, Boston, MA, and Washington, DC, 2009, p. 2.

23. Ibid., p. 5.

24. Information Resources, Inc., "Sustainability: CPG Marketing in a Green World," Information Resources, Inc., Chicago, IL, 2009, p. 7.

25. Ibid., p. 10.

26. Sustainable Packaging Coalition and Packaging and Technology Integrated Solutions, "The Essentials of Sustainable Packaging," Sustainable Packaging Coalition and Packaging and Technology Integrated Solutions, Charlottesville, VA, and Kalamazoo, MI, 2009.

27. Grocery Manufacturers Association and Deloitte Consulting, "Finding the Green in Today's Shopper," p. 7.

28. Ibid., pp. 38–39.

29. Datamonitor, "Sustainable Packaging Trends: Consumer Perspectives and Product Opportunities," Datamonitor, London, 2009, p. 23.

30. Walter Soroka, *Fundamentals of Packaging, Fourth Edition*, p. 49.

31. Grocery Manufacturers Association and Deloitte Consulting, "Finding the Green in Today's Shopper," p. 11–12.

32. John Kalkowski, "Green Is Ingrained in Packaging," in *Packaging Digest* (Oak Brook, IL: Reed Business Information, 2009), p. 24.

33. Associated Press, "Wal-Mart to Cut Packaging by 5 Percent," September 22, 2006.

34. Wal-Mart Stores, Inc., "Wal-Mart Is Taking the Lead on Sustainable Packaging," December 2008, *www.walmartstores.com*.

35. Ibid., p. 2.

36. Erica L. Plambeck, "The Greening of Wal-Mart's Supply Chain," *Supply Chain Management Review* (July 1, 2007).

37. Ibid.

38. Wal-Mart Stores, "Sustainability Packaging."

39. Dana Luthy, Interview, August 2009.

40. Target Corporation, "2009 Corporate Responsibility Report," Target Corporation, St. Paul, MN, 2009, p. 33.

41. Tesco PLC, "Corporate Responsibility Report 2009," Tesco PLC, Cheshunt, UK, 2010, p. 14.

42. Organization for Economic Cooperation and Development (OECD), "Fact Sheet: Extended Producer Responsibility," 2006, *www.oecd.org*.

43. Margaret Walls, "EPR Policies and Product Design: Economic Theory and Selected Case Studies," OECD, Paris, 2006, p. 40.

44. EPA.gov, "About CPG/RMAN," Environmental Protection Agency, Washington, DC, 2008.

45. Garth Hickle, "Product Stewardship Recommendation Report," Minnesota Pollution Control Agency, St. Paul, MN, 2009, p. 3–4.

46. EUROPA European Commission, "Packaging and Packaging Waste," European Union, Web site, 2007.

47. Duales System Deutschland, GmbH, "Company Information/Packaging Ordinance/Prevention, Reduction, Recycling," 2009, *www.gruener-punkt.de.*

48. Ibid.

49. EUROPEN, *Understanding the CEN Standards on Packaging and the Environment: Some Questions and Answers*, 4th edition (Brussels: EUROPEN, 2006), p. 5.

50. European Parliament and Council Directive 94/62/EC of December 20, 1994, on packaging and packaging waste, Annex II.

51. GreenerDesign staff, "Creation of International Packaging Standards Begins," Greenbiz.com, December 10, 2009.

52. EUROPEN, "Why Global Packaging Standards are Urgently Needed," EUROPEN, Brussels, 2009.

53. Dale Andersen, Interview, January 2010.

54. Impact Marketing Consultants, Inc., *The Marketing Guide*, p. 4.

55. Walter Soroka, *Fundamentals of Packaging Technology, Fourth Edition*, p. 495.

56. Ibid., p. 508.

57. John Blanchard and Sal Spada, "Packaging Automation: Centerpiece for CPG Business Strategies," ARC Advisory Group, Boston, MA, 2008.

58. Walter Soroka, *Fundamentals of Packaging Technology, Fourth Edition*, p. 128.

59. Wexxar Packaging, "Wexxar Packaging's new motion-enhanced Pin & Dome System on WF30 case formers increases reliability of using recycled corrugated," press release, Wexxar Packaging, Delta, BC, September 24, 2009.

60. John Blanchard and Sal Spada, "Packaging Automation: Centerpiece for CPG Business Strategies."

61. U.S. Environmental Protection Agency, "2000 Small Business Awards," Green Chemistry, *www.epa.gov/greenchemistry/pubs/pgcc/winners/sba00.html.*

62. John Blanchard and Sal Spada, "Packaging Automation: Centerpiece for CPG Business Strategies."

63. Andrew Joseph, "Straight Down the Line," *Canadian Packaging*, April 2008.

64. Peter Fox, "Quantifying the Environmental Impact of Secondary Packaging," Delkor Systems, Inc., Minneapolis, MN, no date, p. 4.

65. Ibid., p. 5.

66. Ibid., p. 6–9.

67. Andersen, Interview.

68. Andersen, Interview.

69. *Emballage Digest*, "ITW Acquires Hartness," *Emballage Digest, e.bonus service*, Boulogne-Billancourt, France, November 10, 2009.

70. Hartness Sustainability Solutions, "Hartness Automation," 2010, *www.hartness.com.*

71. Hartness Sustainability Solutions, "Hartness Integration," 2010, *www.hartness.com*.

72. Anne-Frances Hutchinson, "Hartness International, Inc.: You Can't Bottle This Kind of Success," *Food & Drink Digest*, Corporate Profile (Norwich, United Kingdom: White Digital Media Ltd./Inc., 2009), p. 5.

73. Scott Smith, Interview, March 2010.

74. Hartness Sustainability Solutions, "Global Messenger" and "FlashBack," 2010, *www.hartness.com*.

75. Smith, Interview, March 2010.

76. Hartness Sustainability Solutions, "Life Cycle Sustainment," 2010, *www.hartness.com*.

77. Hartness Sustainability Solutions, "Hartness DCL," 2010, *www.hartness.com*.

78. Hartness Sustainability Solutions, "Hartness-Inks," 2010, *www.hartness.com*.

79. Hartness Sustainability Solutions, "Grab Pack," 2010, *www.hartness.com*.

80. Hartness International, "Apply the Possibilities," Hartness International, Greenville, SC: no date.

81. Food productiondaily.com, "Hartness and RKW join up to launch new multipack concept" (Montpelier, France: Decision News Media, September 23, 2009).

82. Scott Smith, Interview, March 2010.

83. Ibid.

84. Ibid.

85. Hartness International, "Appointment of new EMEA Managing Director reflects growth plans for Hartness International, Johnson returns to Hartness to support European development," press release, Hartness International, March 2008.

86. *Emballage Digest*, "From decoration to secondary packaging, an original and complete offer," *Emballage Digest* 529 (June 2008).

87. Hutchinson, "You can't bottle this kind of success," p. 3.

88. Nordson Corporation, *Solid Performance. Strongly Positioned. Nordson Corporation 2009 Annual Report* (Westlake, Ohio: Nordson Corporation, 2010), pp. 1, 24, 67.

89. Rick Pallante, "Advances in Adhesive Dispensing Technology Move Packaging to Greater Sustainability," Nordson Corporation, Duluth, GA, 2009.

90. Ibid.

91. Ibid.

92. Ibid.

93. Selim Yalvac and David Mitchell, "Metallocene-based Hot Melt Adhesives for Case and Carton Sealing Applications," white paper, Dow Chemical Company, 2007, p. 2.

94. Pallante, Interview, January 2010.

95. Pallante, "Advances in Adhesive Dispensing Technology," 2009, p. 7.

96. Pallante, Interview, January 2010.

97. Ibid.

98. Ibid.

99. Ibid.

100. Jack Mans, "The game is on the line," *Packaging Digest* (May 1, 2008).

101. Paul Appelbaum, Interview, March 2010.

102. Partner Pak Incorporated, "UV Adhesive/ General Info," Partner Pak Incorporated, 2010, *www.partnerpak.com*.

103. *ThomasNet News,* "Dymax Partners with Partner Pak," Thomas Publishing Co., May 1, 2008.

104. Partner Pak Incorporated, "Packaging Sealing/Comparison UV v RF/Sonic," Partner Pak Incorporated, no date, *www.partnerpak. com.*

105. Paul Appelbaum, Interview, February 2010.

106. Walter Soroka, *Illustrated Glossary of Packaging Terminology* (Naperville, IL: Institute of Packaging Professionals, 2008), p. 18, 33.

107. Partner Pak Incorporated. "Trapped Card v. Trapped Blister," Partner Pak Incorporated, 2010, *www.partnerpak.com.*

108. Mark W. Anderson, "Incorporating Sustainability in Packaging Machine Design," (Cincinatti, Ohio: Pro Mach, Incorporated, undated presentation).

109. Pro Mach Incorporated, "About Pro Mach," Pro Mach Incorporated, Cincinnati, OH, 2010, *www.promachinc.com.*

110. Wexxar Packaging, "Wexxar Packaging's new motion-enhanced Pin & Dome System."

111. Wexxar Packaging, "Wexxar Packaging's new motion-enhanced Pin & Dome System."

112. William Chu, Interview, March 2010.

113. Ibid.

114. Mark W. Anderson, "Incorporating Sustainability in Packaging Machine Design."

115. Ibid.

116. Ibid.

117. Chu, Interview, March 2010.

118. Philip G. Kuehl, *2007 Packaging Machinery Purchasing Process Study* (Arlington, VA: Packaging Machinery Manufacturers Institute, 2007), p. 10-12.

119. Discussion organized by Purdue University Calumet, Department of Mechatronics, July 20, 2009.

120. Philip G. Kuehl, *2007 Packaging Machinery Purchasing Process Study*, p. 3, 10.

121. Purdue University Calumet, 2009.

122. Deloitte Consulting, "Boardroom to Breakroom," p. 26.

123. Ben Miyares, Interview, May 2009.

124. Deloitte Consulting, "Boardroom to Breakroom," p. 26.

125. John Kowal, Interview, September 2009.

126. Anne Johnson, Interview, May 2009.

127. Jack Aguero, Interview, March 2010.

128. Victor Bell, "'Extended Producer Responsibility' Shifts Responsibility for Packaging Disposal to Manufacturers," Environmental Packaging International, Providence, RI, 2010.

129. Jordan Berkley, "Greening Manufacturing Operations and the Supply Chain with Process Efficiency," Apriso Corporation, (Long Beach, CA, 2010.

130. Kevin T. Higgins, "High hopes, low budgets," *Food Engineering* (Troy, MI: BNP Media, February 2010), p. 38.

131. David Dixon, Interview, March 2010.

132. Paul Appelbaum, Interview, February 2010.

133. European Union, Directorate General Environment, "Recast of the WEEE and RoHS Directives proposed," European Union, Directorate General Environment, Brussels, February 2010.

134. EU Directive on Waste Electrical and Electronic Equipment, Annexes 1A and 1B, 2003.

135. Pallante, Interview, January 27, 2010.

136. Deloitte Consulting "Boardroom to Breakroom," p. 27.

137. Packaging Machinery Manufacturers Institute, "PMMI Certified Trainer and Documentation Overview," Packaging Machinery Manufacturers Institute, Arlington, VA, 2008, p. 13.

138. Sustainable Packaging Coalition and Packaging and Technology Integrated Solutions, "The Essentials of Sustainable Packaging," 2009.

139. Consumer Goods Forum, "Retailers and Manufacturers Join Forces to Drive Global Change in Packaging" (Paris: Consumer Goods Forum, January 27, 2010); Sustainable Packaging Coalition, *Sustainable Packaging Indicators and Metrics Framework* (Charlottesville, VA: Sustainable Packaging Coalition, 2009), p. 5; PepsiCo, "Essay: Tom Delay, Chief Executive, The Carbon Trust/ Letter from Tom Delay, Chief Executive, The Carbon Trust," PepsiCo, Purchase, NY, 2009, *www.pepsico.com*; British Standards Institute, "PAS 2050: Specification for the Assessment of the Life Cycle Greenhouse Gas Emissions of Goods and Services," British Standards Institute, London, 2009; Sustainable Packaging Coalition, "The Essentials of Sustainable Packaging," 2009.

140. Carl Wright, "Understanding Overall Equipment Effectiveness," *Reliable Plant Magazine*, 2010, *www.reliableplant.com/ Read/11785/overall-equipment-effectiveness*.

141. Ibid.

142. Ibid.

143. Global Reporting Initiative, "About GRI/ FAQs," Global Reporting Initiative, Amsterdam, 2007, *www.globalreporting.org*.

144. Ibid.

145. Sustainable Packaging Coalition, *Sustainable Packaging Indicators & Metrics, Version 1.0* (Charlottesville, VA: Sustainable Packaging Coalition, 2009).

Appendix 1: Sustainable Packaging Coalition Definition of Sustainable Packaging

- Is beneficial, safe, and healthy for individuals and communities throughout its life cycle

- Meets market criteria for both performance and cost

- Uses renewable energy for sourcing, manufacturing, transporting, and recycling

- Optimizes renewable or recycled source materials

- Applies clean production technologies and best practices

- Uses materials healthy in all probable end-of-life scenarios

- Optimizes materials and energy

- Is effectively recovered and used in biological and industrial closed loop cycles

Appendix 2: Sustainable Design Options for Packaging Machinery

Sustainable Design Technique	Sustainability Savings
Substituting air-actuated devices for mechanical grippers and cylinders	Reduces long-term energy costs, noise pollution, hazardous waste, and maintenance costs; simplifies machinery design
Using electromechanical actuators to divert, lane, collate, and reject packs or products	Reduces reliance on air-actuated devices to save energy and to reduce noise pollution and hazardous waste
Substituting centralized mechanical pumps for venturi-type vacuum pumps in product gripper applications	Reduces the distance of air flow to save energy
Replacing cabinet-mounted drives with integrated motor or drive modules distributed along the machine	Allows for ambient cooling as opposed to control cabinet fan and air conditioner cooling, thereby saving energy
Replacing mechanical drivetrains with multiple continuous motion servo machines	Eliminates the shock of mechanical drivetrains on a main lineshaft that engages and disengages from a large motor, therebysaving energy
Substituting aluminum and composites for steel and cast iron	Reduces inertial loads and thus energy consumption
Using available control technology to tightly synchronize packaging machines	Reduces the energy needed to keep a process line well balanced and running smoothly
Running multiple drives from a shared power supply or common DC bus	Results in energy savings via regeneration
Incorporating faster processors in machine controllers	Uses faster, more responsive processors that allow for smaller motors and lower energy consumption
Replacing zerk fittings with centralized lubrication	Prevents lack of lubrication, which causes friction, requiring additional current to drive the load and thus additional energy use
Incorporating dynamic power monitors that can measure energy usage	Allows users to benchmark and measure the impact of packaging machines on production costs and environmental footprint

Source: Schneider Electric

Appendix 3: Model TCO Calculator (incorporating sustainability features)

Description	Amount
Initial investment	
Equipment base price	$_____
Equipment options	$_____
Factory acceptance test	$_____
Platforms	$_____
Spare parts	$_____
Installation costs	$_____
Training costs	$_____
Training documentation	$_____
Start up costs: technician onsite	$_____
Onsite qualification and verification support	$_____
Service contract	$_____
Warranty (beyond standard period)	$_____
Taxes	$_____
Freight charges	$_____
Annual operating and maintenance costs (to be aggregated over product's expected lifespan)	
Maintenance costs	
Spare parts: electrical	$_____
Spare parts: mechanical	$_____
Technical service labor costs	$_____
Changeover and downtime costs	
Downtime for change parts	$_____
Expected service downtime	$_____
Natural resource consumption and waste disposal costs	
Electricity/energy	$_____
Water	$_____

continued on page 38

continued from page 37

Description	Amount
Line lubrication	$_____
Solid waste generated and handling cost	$_____
Hazardous waste generated and handling cost	$_____
Wastewater generated and handling cost	$_____
Salvage value at end of life	($_____)
Total cost of ownership	$_____

Source: Adapted from materials developed by Deloitte Consulting LLP and Hartness International.

Appendix 4: Contacts

Allied Development
www.allied-dev.com

Delkor Systems, Inc.
www.delkorsystems.com

European Commission
ec.europa.eu/index_en.htm

EU Directive on Packaging and Packaging Waste
*eur-lex.europa.eu/LexUriServ/LexUriServ.do?uri=OJ:
L:2005:070:0017:0018:EN:PDF*

EU Directive on Waste Electrical and Electronic
Equipment
*eur-lex.europa.eu/LexUriServ/LexUriServ.do?uri=OJ:
L:2008:081:0065:0066:EN:PDF*

EU Directive on Restriction of Hazardous Substances
*eur-lex.europa.eu/LexUriServ/LexUriServ.do?uri=OJ:
L:2008:081:0067:0068:EN:PDF*

Hartness International
www.hartness.com

Nordson Corporation
www.nordson.com

Packaging Machinery Manufacturers Institute
www.pmmi.org

Partner Pak, Inc.
www.partnerpak.com

Pro Mach, Inc./Wexxar Packaging
www.wexxar.com

Purdue University-Calumet/Mechatronics Program
*http://webs.calumet.purdue.edu/et/eng-tech/mecha-
tronics-engineering-technology-program-overview/*

Sustainable Packaging Coalition
www.sustainablepackaging.org

U.S. Environmental Protection Agency
www.epa.gov